IET CONTROL ENGINEERING SERIES 73

# Adaptive Sampling with Mobile WSN

**Other volumes in this series:**

Volume 2 **Elevator traffic analysis, design and control, 2nd edition** G.C. Barney and S.M. dos Santos
Volume 8 **A history of control engineering, 1800–1930** S. Bennett
Volume 14 **Optimal relay and saturating control system synthesis** E.P. Ryan
Volume 18 **Applied control theory, 2nd edition** J.R. Leigh
Volume 20 **Design of modern control systems** D.J. Bell, P.A. Cook and N. Munro (Editors)
Volume 28 **Robots and automated manufacture** J. Billingsley (Editor)
Volume 32 **Multivariable control for industrial applications** J. O'Reilly (Editor)
Volume 33 **Temperature measurement and control** J.R. Leigh
Volume 34 **Singular perturbation methodology in control systems** D.S. Naidu
Volume 35 **Implementation of self-tuning controllers** K. Warwick (Editor)
Volume 37 **Industrial digital control systems, 2nd edition** K. Warwick and D. Rees (Editors)
Volume 39 **Continuous time controller design** R. Balasubramanian
Volume 40 **Deterministic control of uncertain systems** A.S.I. Zinober (Editor)
Volume 41 **Computer control of real-time processes** S. Bennett and G.S. Virk (Editors)
Volume 42 **Digital signal processing: principles, devices and applications** N.B. Jones and J.D.McK. Watson (Editors)
Volume 44 **Knowledge-based systems for industrial control** J. McGhee, M.J. Grimble and A. Mowforth (Editors)
Volume 47 **A history of control engineering, 1930–1956** S. Bennett
Volume 49 **Polynomial methods in optimal control and filtering** K.J. Hunt (Editor)
Volume 50 **Programming industrial control systems using IEC 1131-3** R.W. Lewis
Volume 51 **Advanced robotics and intelligent machines** J.O. Gray and D.G. Caldwell (Editors)
Volume 52 **Adaptive prediction and predictive control** P.P. Kanjilal
Volume 53 **Neural network applications in control** G.W. Irwin, K. Warwick and K.J. Hunt (Editors)
Volume 54 **Control engineering solutions: a practical approach** P. Albertos, R. Strietzel and N. Mort (Editors)
Volume 55 **Genetic algorithms in engineering systems** A.M.S. Zalzala and P.J. Fleming (Editors)
Volume 56 **Symbolic methods in control system analysis and design** N. Munro (Editor)
Volume 57 **Flight control systems** R.W. Pratt (Editor)
Volume 58 **Power-plant control and instrumentation** D. Lindsley
Volume 59 **Modelling control systems using IEC 61499** R. Lewis
Volume 60 **People in control: human factors in control room design** J. Noyes and M. Bransby (Editors)
Volume 61 **Nonlinear predictive control: theory and practice** B. Kouvaritakis and M. Cannon (Editors)
Volume 62 **Active sound and vibration control** M.O. Tokhi and S.M. Veres
Volume 63 **Stepping motors: a guide to theory and practice, 4th edition** P.P. Acarnley
Volume 64 **Control theory, 2nd edition** J.R. Leigh
Volume 65 **Modelling and parameter estimation of dynamic systems** J.R. Raol, G. Girija and J. Singh
Volume 66 **Variable structure systems: from principles to implementation** A. Sabanovic, L. Fridman and S. Spurgeon (Editors)
Volume 67 **Motion vision: design of compact motion sensing solution for autonomous systems** J. Kolodko and L. Vlacic
Volume 68 **Flexible robot manipulators: modelling, simulation and control** M.O. Tokhi and A.K.M. Azad (Editors)
Volume 69 **Advances in unmanned marine vehicles** G. Roberts and R. Sutton (Editors)
Volume 70 **Intelligent control systems using computational intelligence techniques** A. Ruano (Editor)
Volume 71 **Advances in cognitive systems** S. Nefti and J. Gray (Editors)
Volume 73 **Adaptive sampling with mobile WSN** K. Sreenath, M.F. Mysorewala, D.O. Popa and F.L. Lewis
Volume 74 **Eigenstructure control algorithms: applications to aircraft/rotorcraft handling qualities** S. Srinathkumar

# Adaptive Sampling with Mobile WSN

## Simultaneous robot localisation and mapping of paramagnetic spatio-temporal fields

Koushil Sreenath,
Muhammad F. Mysorewala,
Dan O. Popa and Frank L. Lewis

The Institution of Engineering and Technology

Published by The Institution of Engineering and Technology, London, United Kingdom

The Institution of Engineering and Technology is registered as a Charity in England & Wales (no. 211014) and Scotland (no. SC038698).



First published 2011



The Institution of Engineering and Technology
Michael Faraday House
Six Hills Way, Stevenage
Herts, SG1 2AY, United Kingdom

www.theiet.org



**British Library Cataloguing in Publication Data**
A catalogue record for this product is available from the British Library

**ISBN 978-1-84919-257-6 (hardback)**
**ISBN 978-1-84919-258-3 (PDF)**

Typeset in India by MPS Ltd, a Macmillan Company
Printed in the UK by CPI Antony Rowe, Chippenham, Wiltshire

# Contents

# Preface

The objective of this book is to present systematic methods for estimating environmental fields using multiple mobile sensors. Monitoring environmental fields is a complex task and is of great use in many areas, such as for building models of natural phenomenon: e.g. agriculture monitoring, such as monitoring soil temperature to manage frost, wind, water, disease and pests; ocean, river and lake monitoring of environmental phenomena such as salinity in lakes, tracking water temperature, particulate densities and pollutants responsible for sustaining marine colonies, or coral cover of oceanic reefs; meteorology monitoring, such as tracking of storms, gas plumes and air quality; forest monitoring for tracking humidity in forests, and prediction and decision making during forest firefighting, etc.

Sampling is a broad methodology for gathering statistical information about a phenomenon. The capabilities and distributed nature of wireless sensor networks provide an attractive sampling approach for estimation of spatio-temporally distributed environmental fields. This is adaptive sampling, where the strategy for 'where to sample next' evolves temporally with past measurements. Thus, the sensor network physically adapts with past measurements to enable sampling at locations that give maximal information about the field being estimated. This book presents adaptive sampling strategies with multiple, heterogeneous and mobile sensors. Sensors of this kind present several complexities, some of which like deadlocks and localization issues are also addressed here.

The book is organized into three parts: preliminaries, single-robot adaptive sampling and multi-resource strategies. The preliminaries begin with Chapter 1, which presents the motivation and introduces the problem to be solved. Chapter 2 introduces two separate test beds that provide means of experimentally validating the theory developed in the book. Part II begins with Chapter 3 on adaptive sampling of fields that can be expressed as parametric models, and the problem reduces to estimating these field parameters. Chapter 4 applies techniques developed in the previous chapter to map forest fires. Part III develops techniques to extend the adaptive sampling strategies developed for single robots onto multiple heterogeneous mobile sensors. Chapter 5 develops strategies to distribute the adaptive sampling processing onto multiple mobile sensors. Chapter 6 presents a control scheme to address the issue of deadlocks that arise when multiple resources need to be allocated for sensing tasks. Finally, Chapter 7 presents localization algorithms in which the static sensor network and the mobile sensors cooperate to simultaneously and adaptively localize themselves.

The algorithms presented in this book arose as ideas by two of the authors' advisers. Their plentiful ideas and confidence in us gradually helped in establishing this work. This book is geared to serve as a source of comprehensive information for researchers, practitioners and graduate students working on mobile sensor networks for building models of environmental phenomena.

The authors wish to thank the Automation & Robotics Research Institute (ARRI), in general, and Prof. Harry E. Stephanou, ARRI director, in particular, for their generous support during the course of this research. We would also like to acknowledge former collaborators at Rensselaer Polytechnic Institute (RPI), Prof. Arthur Sanderson, Vadiraj Hombal, and Richard Blidberg, director of the Autonomous Undersea Institute (AUSI). Adaptive sampling in environmental monitoring using information measures was the first application of our research when our research started in 2003.

We would also like to thank Dr Vincenzo Giordano and Dr Prasanna Ballal for establishing the mobile wireless sensor network test bed, pioneering several initial results and further extending some of the deadlock-based work presented here; Ms Jaymala Ghadigaonkar and Mr Rohit Talati for contributions in building a revised and more robust version of the ARRI-Bots; King Fahd University of Petroleum and Minerals, Saudi Arabia, for currently supporting research in this area, and its application to real problems; and finally Dr Nigel Hollingworth, editor, *IET*, and his team for being flexible to let us work around our schedules and for their swift and professional handling of the entire publication process.

Work presented here was supported in part by the ARRI, the Army Research Office grants W91NF-05-1-0314 and M-47928-CI-RIP-05075-1, and the National Science Foundation grants IIS-0326595 and CNS-0421282.

Koushil Sreenath, Ann Arbor, Michigan
Muhammad F. Mysorewala, Dhahran, Saudi Arabia
Dan O. Popa, Arlington, Texas
Frank L. Lewis, Arlington, Texas
May 2010

*Part I*

# Preliminaries

# *Chapter 1*
# Introduction

Mobile robots are being increasingly used on land, underwater and in air, as sensor-carrying agents to perform sampling missions, such as searching for harmful biological and chemical agents, search and rescue in disaster areas, and environmental mapping and monitoring. Sampling problems are central aspects of deployment because complete coverage is not possible if the environment is large and has only a few ‘hot spots’, or if the sampling costs are high. The capabilities and distributed nature of wireless sensor networks (WSNs) then provide an attractive sampling approach for estimation of spatio-temporally distributed environmental phenomena. Using a densely deployed static sensor network to cover large sampling volumes serves as a solution to the sampling problem; however, it is very expensive in time and resource costs and places heavy demands on energy consumption. Thus, physical adaptation of a sensor network, either by adaptive sensor scheduling or through robotic mobility, may be the only practical approach. The spatio-temporally varying network topologies of mobile sensor networks make them suitable for the task of reconstruction of a parametric environmental field through adaptive sampling (AS) at locations that maximally reduce the uncertainty in the field parameter estimates. This leads to AS wherein sampling strategies temporally evolve with past measurements.

Sensor networks are composed of a mix of both static and mobile sensors. The large scale of deployment in sensor networks makes careful placement or uniform distribution of sensor nodes impractical, and static sensor nodes end up being deployed with no prior knowledge of their location. However, location information is imperative in sensor networks for most applications for localized sensing. Thus, localizing the sensor network adaptively with no additional hardware becomes necessary. This leads to a simultaneous adaptive localization wherein sensor network localization occurs in such a way to maximally reduce uncertainty in localization of the sensor network.

Mobile sensor networks are composed of multiple heterogeneous resources, where multiple tasks contend for a single shared resource, and multiple resources contend to perform a single task. In the former case, we have shared resources, and in the latter case, we have routing resources. Thus, a method of deadlock-free resource scheduling in the presence of shared and routing resources becomes necessary to schedule the most efficient (cost/energy/time) resource for a task. This leads to a discrete event controller (DEC) to achieve a deadlock-free resource allocation.

In the following sections, a broad overview of the three central areas of the book – adaptive sampling, adaptive localization and deadlock-free resource scheduling – is presented in the context of mobile WSNs.

## 1.1 Adaptive sampling for density estimation using WSN

The problem of environmental field estimation can be addressed through spatio-temporal measurements using a WSN. A WSN that provides proper coverage of the field to be estimated would easily provide sufficient data to solve the problem of estimating a field distribution. However, as briefly eluded, complete coverage of an area is not possible if the environment is large, has only a few 'hot spots' or if the sampling costs are high. Although using a densely deployed static sensor network to cover large sampling volumes serves as a solution, it is extremely expensive in terms of time, resource and energy costs. Several researchers have worked on increasing the coverage through innovative methods.

To quantify coverage, methods to determine whether every point in the field area is covered by at least *k* sensors are formulated [1]. Various methods of sensor node positioning to provide good coverage are discussed in Reference 2. As an example of sensor node positioning developed in Reference 3, the sensor nodes are placed such that they provide coverage that guarantees detection of an intruder passing through a border. With a dense sensor node deployment, the lifetime of the sensor network can be increased by scheduling selected sensors into an inactive state while still maintaining coverage, as illustrated by various approaches in References 4 and 5. Another method of increasing sensor network coverage is illustrated in Reference 6, where noisy data from multiple sensors is fused for reconstructing measurements from areas that are not truly covered by any single sensor. The effect of various environmental factors on coverage is studied in Reference 7. Instead of having complete coverage through densely deployed static sensors, mobile sensors can be used. Although mobile sensor networks do not provide complete coverage at any given instant of time, a maximal detection time can be formulated, which guarantees that a particular location is sensed within this time [8]. Mobility of the sensor nodes can also be used to improve connectivity and data throughput in a sensor network [9]. Thus, mobile sensors offer an effective way to sense the field without having to resort to dense sensor deployments. Then, the problem of where to sample next using the mobile sensor arises and leads to AS.

In the context of optimizing the sampling of spatio-temporal fields with mobile robots, several AS algorithms have recently been proposed [10–14]. The term 'adaptive' refers to choosing the sampling points based on the amount of information they provide about the spatio-temporal distribution being mapped. Examples of spatially distributed fields that can be monitored using AS are salinity in lakes [13], humidity in forests [14] and chemical leaks in buildings, as is done in odour sampling [10,15,16].

Sampling using mobile robots is a typical example of a real-time density estimation problem and has been covered in detail in past work [17,18]. The sensor fusion approaches for density estimation can be classified into three different categories. The first is building physical parametric models, the second is parametric and non-parametric feature-based inference, both of which involve clustering of observations, and the third is cognitive-based modelling.

In general, a spatio-temporal distribution can be modelled in a parametric or non-parametric manner. Strictly non-parametric field descriptions are accurate, as they do not assume any functional form of distribution in advance. However, they require a lot of samples to reconstruct the distribution, which seems difficult to apply in practice to situations involving time and energy constraints, and for time-varying fields. An example of solutions with non-parametric distribution assumptions is discrete landmark mapping [19] that involves exploration, instead of planned sampling missions. Other examples are chemical source localization [16] using *hill* (or *gradient*) *climbing approach* and *adaptive cluster sampling* [20] for wide-area sampling.

In contrast, parametric algorithms are computationally efficient and fairly easy to run onboard mobile robots in a distributed fashion. Such models can sometimes be constructed from physical or numerical models such as in the case of weather prediction. With no *a priori* information, the sampling problem becomes an exploration problem in the beginning, until enough samples are taken and distribution assumptions can be made. The advantage of the parametric model is that, in fact, it can also include a parametric model of sensor observable data and corresponding measurement uncertainties. Also, the uncertainty in the field estimate can be measured indirectly through uncertainty in the distribution parameters. Furthermore, parametric modelling works better for time-varying fields, in which the field variation can be represented through parameters instead of the entire density function. The main limitation of parametric representations is modelling uncertainty that adds another source of estimation error.

A problem that naturally arises in the process of sampling is making sure that the measurements of sensors are correlated with their position, and that the data from multiple sensors is fused efficiently. Multiple vehicle localization and sensor fusion are by now classic problems in robotics, and there has been considerable progress in the past two decades in these areas [21,22]. Furthermore, distributed field variable estimation is relevant to charting and prediction in oceanography and meteorology, and has also received considerable attention [23]. In both contexts, measurement uncertainty can be addressed using Kalman filter estimation [24]. Recently, Popa *et al.* described a combined multi-agent AS problem coupling wireless sensor nodes with mobile robots and using information measures to reposition the robots in order to achieve near-optimal sampling of a distributed field [13]. Unlike other non-parametric sampling methods [11,14], this approach requires a parametric field description of the sampled field, and a dynamic model for robots. The advantage of this approach was the inclusion of vehicle localization uncertainty, but the disadvantage was the fact that it could only handle very simple field models.

Localization uncertainties are especially relevant for robotic vehicles where dead-reckoning errors are high such as in the case of global positioning system (GPS)-denied environments. The field model parameters are combined with uncertainty in the estimation of the robot localization using the overall error covariance. Therefore, localization uncertainty is reduced by building accurate models of distributed fields and vice versa. For instance, if a robot is sampling an unknown field, but its location is accurately known, a distributed parameter field model can be constructed by taking repeated field samples. Later on, this field model can be used to reduce the robot localization error.

In this book, we follow a similar parametric approach and use sensor measurements to improve both the field estimate and the localization of the robot. Since this kind of formulation requires a parametric model for a potentially unknown and complex field, we considerably extend and expand on results from prior work. In particular, we address challenges related to sampling of complex, non-linear and time-varying space–time fields and investigate considerations of computational complexity, communication overhead, time and energy constraints, and sampling using multiple robots.

## 1.2 Simultaneous adaptive localization

Sensor networks are composed of a mix of both static sensors and mobile sensors. The large scale of deployment in sensor networks makes careful placement or uniform distribution of static sensor nodes impractical, and usually the sensor nodes are deployed with no prior knowledge of their location. However, location information is imperative for applications in both WSNs and mobile robotics. Many sensor network applications, such as tracking targets, environmental monitoring and geospatial packet routing, require that the sensor nodes know their locations. The requirement of the sensors to be small, untethered, low energy consuming, cheap, etc., makes the sensors resource-constrained [25], and requiring every sensor to be GPS enabled is impractical. Thus, localization is a challenging problem and yet crucial for many applications.

Approaches to the problem of localization are varied. A detailed introduction to localization in sensor networks is presented in Reference 26. GPS [27] solves the problem trivially, but equipping the sensors with the required hardware may be impractical. A small section of active beacons can be placed in the sensor network, and other sensors can derive their location from these anchor nodes [28,29]. Cooperative localization methods have been developed for relative localization [30,31]. Other approaches involve received signal strength indicator (RSSI) [32], time of arrival (TOA) [33,34], Angle of Arrival (AOA) [35] and signal pattern matching [27]. An overview of various methods for localization in sensor networks is presented in Reference 36, and a comprehensive coverage of topics and fundamental theories behind localization algorithms in WSN is developed in Reference 37.

For localization with no additional hardware on the sensor node, the geometric constraints of radio connectivity are exploited. Some authors suggest using a mobile robot (whose position is known) to localize the sensors. However, the position of the mobile robot may be hard to determine. Sequential Monte Carlo localization [38] presents a range-free localization method in the presence of mobility using mobile and seed nodes. LaSLAT [39] uses a Bayesian filter to localize the sensor network and track the mobile robot. In Reference 40, a particle filter is employed to localize elements of the network based on observation of other elements of the network. In Reference 41, a mobile robotic sensor localized the network based on simple intersections of bounding boxes. In Reference 42, geometric constraints based on both radio connectivity and sensing of a moving beacon localize the sensor network. The Kalman filter has been used in dead reckoning for mobile robots, but its full potential in localization of WSN has not heretofore been fully explored. In Reference 43, an extended Kalman filter (EKF) is used for localization and tracking of a target. Reference 44 uses RSSI and a robust EKF implemented on a mobile robot to localize a delay-tolerant sensor network. The Kalman filter was used in Reference 45 for active beacon and mobile autonomous underwater vehicle (AUV) localization and in Reference 46 for scheduling of sensors for target tracking. Simultaneous localization and mapping (SLAM) [47] and concurrent mapping and localization (CML) [48] employ Kalman filters for concurrent mapping and mobile robot localization, which can be considered similar to our work wherein the geometric constraints introduced due to radio connectivity of the static sensors play the role of features. In this book, we use the full capabilities of the Kalman filter in the general WSN localization problem. A comprehensive development of SLAM is carried out in Reference 49.

Our work exploits geometric constraints based on radio connectivity such that range information is not needed. A mobile robot initially sweeps the network, and broadcasts from the robot are used to localize the sensor nodes. Computationally inexpensive Kalman filters implemented on the sensors fuse the information. On the other hand, as time passes, the mobile robot gradually loses its own localization information. We present an algorithm that uses updates from the better localized sensors along with GPS updates, when they occur, to correct this problem. A continuous–discrete extended Kalman filter (CD EKF) running on the robot estimates the robot state continuously and fuses the discrete measurement updates.

Finally, an adaptive localization algorithm, based on AS techniques [13,50], is presented that navigates the mobile robot to an area of nodes with highest position uncertainty. This ensures that the robot manoeuvres to an area where the nodes are least localized, so that it can maximize the usefulness of its positional information in best localizing the overall network. The adaptive localization strategy ensures that, with a minimal robot movement, the largest reduction in aggregated node uncertainty is achieved at every iteration of the adaptive localization algorithm.

## 1.3 Discrete event controller for resource scheduling

Mobile WSNs are composed of multiple heterogeneous resources capable of performing diverse tasks such as measuring, manipulating, moving and sensing. In mobile sensor networks, a strong one-to-many mapping exists between a resource and the tasks that the resource can perform. This mapping can be statically assigned resulting in shared resources, or dynamically assigned resulting in both shared and routing resources. Shared resources arise when multiple tasks contend for a single shared resource, while routing resources arise when multiple resources contend to perform a single task. The use of shared or routing resources is a major problem occurring in discrete event (DE) systems, including manufacturing systems, computer systems, communication systems, highway/vehicle systems and others [51]. Failure to suitably assign, dispatch or schedule resources in the presence of shared or routing resources can cause serious deleterious effects on system performance, resulting in extreme cases in system deadlock. The need then arises for deadlock prevention, deadlock avoidance or deadlock detection and recovery.

Deadlock avoidance algorithms have been used in various scenarios such as robotic cells [52,53], e-commerce-driven manufacturing systems [54], process control such as semiconductor fabrication [55], communication network routing [56] and computer operating systems. The implementation of deadlock avoidance policies in autonomous distributed robotic systems such as mobile sensor networks has not been still thoroughly investigated. Preliminary simulations of efficient deadlock avoidance policies for shared resources in heterogeneous mobile sensor networks are presented in Reference 57.

A large amount of research has been done in developing various deadlock avoidance algorithms using varied concepts such as circular wait (CW), circular blocking, siphons in Petri nets and critical subsystems. Petri net–based deadlock prevention polices [58,59] deal with detecting siphons and statically introducing control places into the net to eliminate unmarked siphons signifying deadlock. In Reference 54, potential deadlock patterns are acquired from an offline simulation of the part processing sequence, and then an online matching/reordering process is made use of to keep the current system state dissimilar to the acquired deadlock patterns. Mathematical formulations of deadlocks and traps by calculation of $s$-invariants of marked graphs using linear algebra are thoroughly discussed in Reference 60. Supervisory control of Petri nets [61] introduces an approach of keeping a Petri net from starvation by using online routing functions instead of traditional offline control places, where the routing function assigns a non-shared resource to perform the task from within a pool of resources. Detailed mathematical analysis of deadlocks and an efficient dispatching policy for deadlock avoidance based on the generalized Kanban scheme using a matrix model for DE systems are presented in References 51–53 and 62–64.

Because of the heterogeneous nature of mobile sensor networks, resources are capable of performing multiple jobs. These are systems with flexible routing where tasks can choose from a set of resources. In such systems with flexible routing, route enumeration can be of exponential complexity and execution of deadlock avoidance constraints is rendered computationally intractable [65]. In References 65 and 66, a control model is developed that allows for small, quickly enumerable subset of less dense routes, which allows for several processing alternatives (routes) at each step while still maintaining deadlock-free operation. In Reference 67, several novel mathematical formulations are constructed for detecting active CWs leading to a deadlock in flexible routing systems; however, no deadlock avoidance algorithm is arrived at.

In this book, we extend the preliminary analysis of deadlock avoidance polices for shared resources in heterogeneous mobile sensor networks to more complicated scenarios. We show through experimental implementation on an actual mobile sensor network test bed the feasibility and effectiveness of the proposed deadlock-free supervisory control in performing complex and simultaneous sequencing of interconnected tasks. Further, a general deadlock avoidance policy for systems with flexible routing, where both shared and routing resources are present, is mathematically formulated and various simulations performed to validate deadlock-free operation.

## 1.4 Summary

This chapter has presented a broad overview of the three central areas of the book – adaptive sampling, adaptive localization and deadlock-free resource scheduling – all in the context of mobile WSNs. In summary, this book makes the following contributions.

An AS framework is developed for combining measurements arriving from mobile robotic sensors of different scales, rates and accuracies, in order to reconstruct a parametric spatio-temporal field. This novel AS algorithm responds to real-time measurements by continuously directing robots to locations most likely to yield maximum information about the sensed field. In addition, the localization uncertainty of the robots is minimized by combining the location states and field parameters in a joint Kalman filter formulation.

A simultaneous localization algorithm is developed and simulated for localization of a sensor network using geometric constraints of radio connectivity. An adaptive localization algorithm is developed to adaptively navigate a mobile robot such that it optimally minimizes the largest localization uncertainty of a sensor network.

Deadlock avoidance techniques developed using the DEC are extended and implemented on a mobile WSN composed of Cybermotion SR2 patrol robots and Berkley motes, such that smooth, deadlock-free resource scheduling occurs in the presence of shared resources. Further, a general mathematical formulation is developed for deadlock avoidance in systems with flexible routing, where both shared and routing resources exist. Simulations are done to validate deadlock-free operation.

The remainder of the book is structured as follows: Chapter 2 introduces two separate test beds that provide means of experimentally validating the theory developed in the book. Part II begins with Chapter 3 on adaptive sampling of fields that can be expressed as parametric models, and the problem reduces to estimating these field parameters. Chapter 4 applies techniques developed in the previous chapter to map forest fires. Part III develops techniques to extend the AS strategies developed for single robots onto multiple heterogeneous mobile sensors. Chapter 5 develops strategies to distribute the AS processing onto multiple mobile sensors. Chapter 6 presents a control scheme to address the issue of deadlocks that arise when multiple resources need to be allocated for sensing tasks. Finally, Chapter 7 presents localization algorithms in which the static sensor network and the mobile sensors cooperate to simultaneously and adaptively localize themselves.

# *Chapter 2*
# Test beds for theory

## 2.1 Adaptive sampling test bed

### *2.1.1 Objectives of the test bed*

The mobile sensor network for adaptive sampling (AS) is set up as illustrated in Figure 2.1. The primary objectives of the test bed are as follows:

- Develop low-cost mobile wireless sensor nodes (MWSN). These nodes are robotic platforms delivering a sensor pack to a location of interest in order to take samples and transmit sensor information through a wireless network.
- Provide means of experimentally validating AS algorithms developed to optimally estimate an environmental field.
- Provide means of validating various localization algorithms, such as the simultaneous adaptive localization algorithms developed in Chapter 7.
- Provide means of energy harvesting of naturally available energy sources such as solar and piezoelectricity for prolonging the sensor life between recharges.

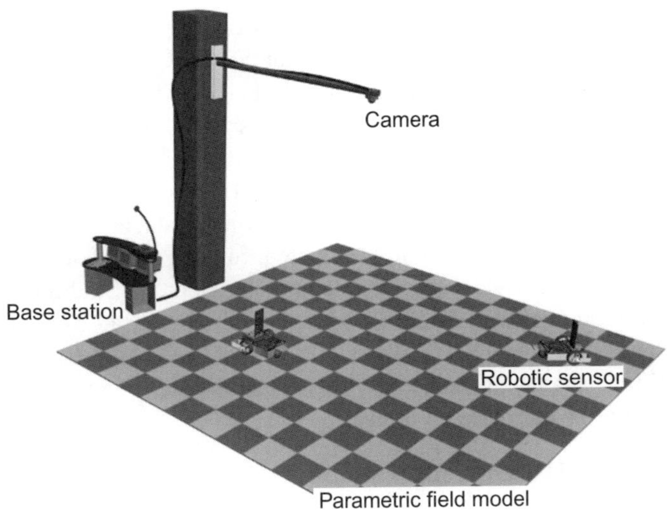


*Figure 2.1 Adaptive sampling test bed*

### 2.1.2 Structure of the test bed

The test bed is composed of multiple MWSN that enable sensing the environment at various locations. An inexpensive overhead camera serves as an indoor GPS offering infrequent location updates to the mobile sensors. These updates can be fused with the robot's internal location estimates computed via various Kalman filter formulations. A parametric colour field on the ground serves as the environmental field to be estimated. This could be either static as a printed colour field or dynamic as a time-varying field projected from an overhead projector. These dynamic fields could be used to simulate the spread of a forest fire, for instance. Finally, a base station serves as a centralized controller that communicates with all the sensors, captures field samples from various locations and builds an estimate of the field. The experimental setup is illustrated in Figure 2.2 (see p. C1). The following sections provide additional details on the various components of the test bed.

#### 2.1.2.1 Parametric colour field

The static parametric field is a printout of a simple linear colour field on the laboratory floor. The field was physically distributed over a search space of 3.15 m × 2.25 m. The camera at a height of 10.5 m encompasses the entire area in its field of view. Three separate linear field models, one for each primary colour (RGB), are used to generate the colour field. The field is given by

$$R = r_0 + r_1 x + r_2 y, \quad G = g_0 + g_1 x + g_2 y, \quad B = b_0 + b_1 x + b_2 y \tag{2.1}$$

with the following numerical values:

$$\begin{array}{lll} r_0 = 0.2307, & r_1 = 0.0012, & r_2 = -0.00048 \\ g_0 = 0, & g_1 = 0.0002, & g_2 = 0.0018 \\ b_0 = 1.0, & b_1 = -0.00078, & b_2 = -0.001 \end{array} \tag{2.2}$$

The sampling mission for the mobile robots, equipped with a colour sensor, is to recover the unknown (to the robot) field coefficients by sampling.

In order to test the sampling algorithms for complex, time-varying, non-linear fields, the printed colour field on the laboratory floor is no longer appropriate. Instead, a camera–projector system is installed as shown in Figure 2.3. The hardware comprises a projector mounted 6.7 m high in the ceiling. The projected image size is 124 in × 93 in, and the image resolution is 1024 × 768 pixels. The projector has a brightness of 3000 lumen so as to perform satisfactorily in a brightly lit laboratory. It is connected to a computer system running MATLAB that generates complex time-varying fields onto the laboratory floor. In addition, a custom-developed infrared (IR) camera serves to replace the overhead camera to provide infrequent localization updates to the mobile robots. (The earlier camera could not be used in this time-varying field setup due to the inability of accurate localization because of the brightly projected image.)

#### 2.1.2.2 Mobile wireless sensor nodes (ARRI-Bots)

ARRI-Bots are MWSN used for experimental validation of AS algorithms and have been entirely custom built at the Automation & Robotics Research Institute (ARRI). Figure 2.4 illustrates the progression of the design of the ARRI-Bot from

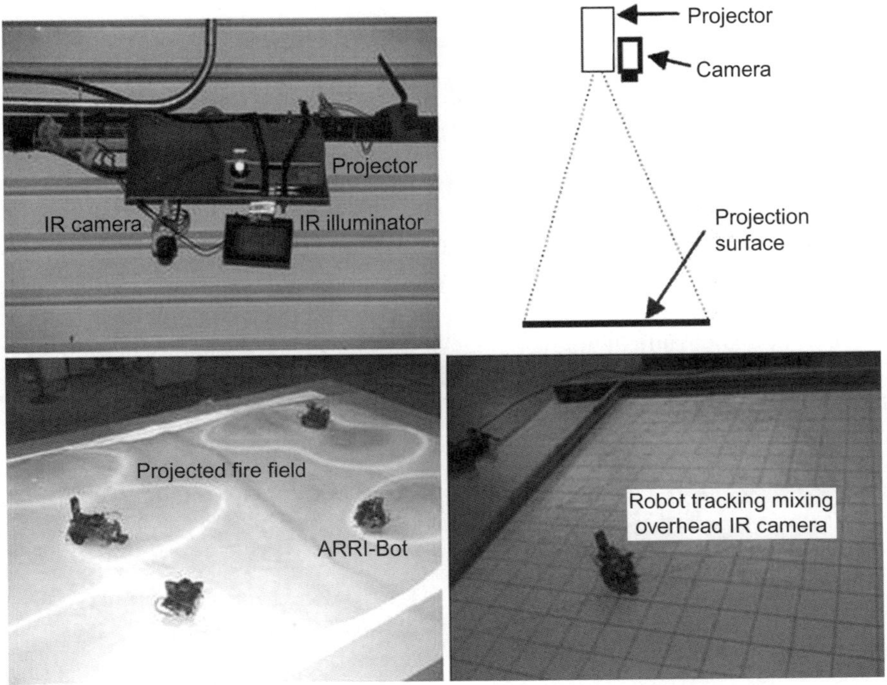


*Figure 2.3 Test bed with time-varying fire field projected on the floor from a projector. ARRI-Bots are shown sampling at various locations for estimating field parameters. Overhead IR camera is used to aid in localization and for validating the accuracy of estimated location*

the prototype stage to the final version (V-2). The ARRI-Bot energy harvester consists of a solar panel and associated electronics for outdoor operation and a piezoelectric energy harvester that augments an onboard Ni-CAD battery pack.

The ARRI-Bot is composed of the following subsystems:

(i) Mechanical subsystem, consisting of an aluminium frame chassis, a differential drive and a front omnidirectional wheel.
(ii) Sensor subsystem, consisting of ultrasonic rangefinders for obstacle detection, colour sensor for sensing the colour field and wheel encoders for dead-reckoning localization.
(iii) Communication subsystem, consisting of a Parallax Transceiver (for ARRI-Bot V-1) or a Cricket® processor/radio module (for ARRI-Bot V-2), operating in the 433-MHz frequency band. Each Cricket Mote is equipped with a default sensor pack (light, temperature, etc.). The Cricket unit carries ultrasonic transducers for range finding and is used for both communication and localization.
(iv) Electronics subsystem, consisting of a custom-built printed circuit board (PCB) board containing a general-purpose Javelin Stamp central processing unit (CPU), a dedicated peripheral interface controller (PIC) microcontroller and glue electronics for interfacing to all the sensor and wireless subsystems. In

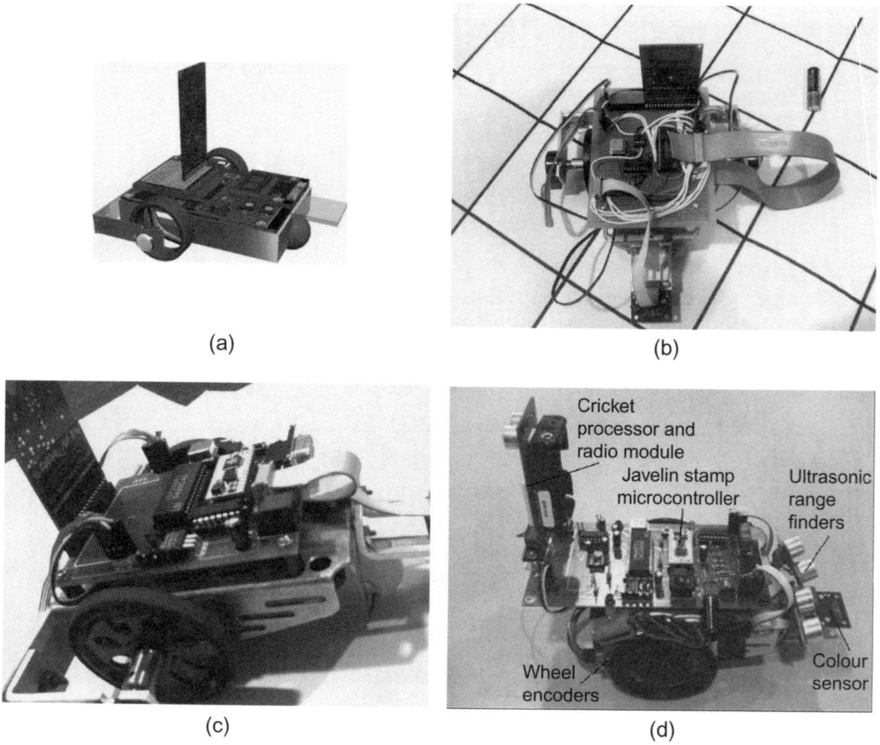


*Figure 2.4 Development of inexpensive ARRI-Bots: (a) initial computer aided design (CAD) model, (b) early prototype, (c) version 1 (V-1) and (d) version 2 (V-2)*

addition, the electronics board contains power management and energy harvesting circuitry.

(v) Power subsystem, consisting of an Ni-MH battery pack, a solar panel for outdoor recharging and operation, and a piezoelectric cantilever array for battery augmentation.

#### 2.1.2.3 Software

Software interfaces are constructed to transparently pass commands and messages between (a) the MATLAB system running the AS algorithms and the base station, (b) the base station and the multiple mobile robotic sensors, and (c) the mobile robot and the onboard sensors.

High-level commands on the ARRI-Bot are implemented to encapsulate actions to be performed by the robot. Commands are simple strings terminated with a terminal character such as the semicolon. These commands achieve motion, sensing and other miscellaneous system tasks. Motion commands are either open loop where the robot is in motion until a stop command is received or closed loop where the robot is in motion until the encoders register a movement corresponding to the commanded amount. Low-level algorithms onboard the mobile robotic sensors perform these commanded actions. Some of the commands are described in Table 2.1.

*Table 2.1 Robot commands*

| Command | Command description |
|---|---|
| F | Move forward |
| B | Move backward |
| R | Turn right |
| L | Turn left |
| S | Stop |
| M 120 | Move forward by 120 encoder counts |
| T-85 | Turn clockwise by 85 encoder counts |
| T 40 | Turn anticlockwise by 40 encoder counts |
| C | Take colour sample (responds with colour read) |
| J | Switch crickets from radio to beacon mode for 15 s |
| Z | Switch crickets from radio to listener mode for 15 s |

To enable implementing even a basic Kalman filter, one needs to perform floating point arithmetic. However, since the onboard microcontroller does not possess a floating point unit (FPU), a fixed-point math library is implemented in software.

## *2.1.3 Calibration algorithms*

The ARRI-Bot is a custom-developed mobile robotic sensor. Thus, one needs to calibrate various parts of the robot to convert from the internal quantities to external physical quantities. For instance, the internal encoder values need to be converted to physical distance moved, overhead camera pixel images need to be converted to physical locations on the parametric field and triangulation distance from beacon Crickets needs to be converted to physical locations on the parametric field. This section discusses these calibration procedures.

### 2.1.3.1 Robot model and dead reckoning

Practical inaccuracies in construction and mechanical assembly, such as unequal wheel radii or incorrect axle length, always exist. This leads to navigational errors in localization. To overcome this, a systematic error [68] is injected into the differential kinematic robot model as indicated in Figure 2.5.

The robot runs an onboard dead-reckoning location estimator using encoder data and nominal physical dimensions (lengths, wheel radii, etc.). The states of the robot are the position and orientation $X = [x \ y \ \theta]^{\mathrm{T}}$. A discrete-time position estimator using the encoder counts $\Delta\varphi_L$ and $\Delta\varphi_R$ is shown in the following equation, where $K_{drv}$ and $K_{turn}$ are the actuator drive constants for either actuator in terms of distance, turn per drive count:

$$\begin{aligned}
\hat{X}_1(k) &= \hat{X}_1(k-1) + \frac{K_{drv-R}\Delta\varphi_R r_R + K_{drv-L}\Delta\varphi_L r_L}{2}\cos(\hat{X}_3(k-1)) \\
\hat{X}_2(k) &= \hat{X}_2(k-1) + \frac{K_{drv-R}\Delta\varphi_R r_R + K_{drv-L}\Delta\varphi_L r_L}{2}\sin(\hat{X}_3(k-1)) \\
\hat{X}_3(k) &= \hat{X}_3(k-1) + \frac{K_{turn-R}\Delta\varphi_R r_R - K_{turn-L}\Delta\varphi_L r_L}{L_b} \\
X &= [\hat{X}_1 \ \hat{X}_2 \ \hat{X}_3] = [x \ y \ \theta]
\end{aligned} \tag{2.3}$$

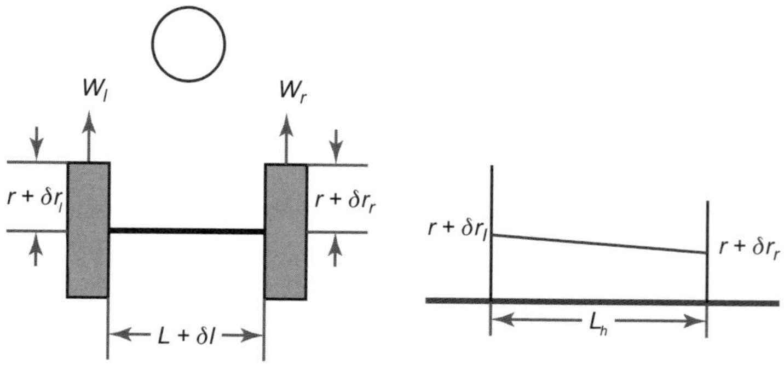


*Figure 2.5 Differential robot with uncertainty in wheel radii and axle length*

A series of Monte Carlo experiments are carried out to identify the constants in the above equation. Then, a simple quasi-holonomic path planner is used to move to a particular location by first turning in place towards the target location and then moving the robot to the target location.

The UMBmark test [69,70], which is a bidirectional square path test, was performed for measuring the dead-reckoning error of the ARRI-Bot. The results shown in Figure 2.6 clearly indicate that the true robot position deviates more from the desired path than the dead-reckoning position. The former involves both systematic and non-systematic influences, whereas the latter can only capture systematic errors.

#### 2.1.3.2 Localization using overhead camera

The overhead camera system serves as a positioning system with low sampling rate in order to correct dead-reckoning navigation errors. Unlike the dead-reckoning error, localization error using vision does not grow with time.

Robot localization using vision is a simple object recognition problem. First, the camera is calibrated to account for barrel distortion that arises due to the wide-angle lens. Next, image segmentation is performed, and the location and orientation are estimated from the triangular shape of the robot. The orientation of the robots can be estimated by using the robot's triangular shape. Figure 2.7 illustrates the steps involved in estimating the location and orientation from an image.

#### 2.1.3.3 Localization using Crickets

The most common way to use Cricket® Motes is to deploy actively transmitting beacons on walls or ceilings and attach listeners to host devices whose location needs to be obtained. As shown in Figure 2.8, crickets on the robot that needs to be localized are placed temporarily in 'listener' mode. Each beacon periodically and simultaneously broadcasts its space identifier and position coordinates on an RF frequency channel and an ultrasonic pulse. Listeners within radio range, ultrasonic range and in line of sight to the beacon receive both messages. Then from the

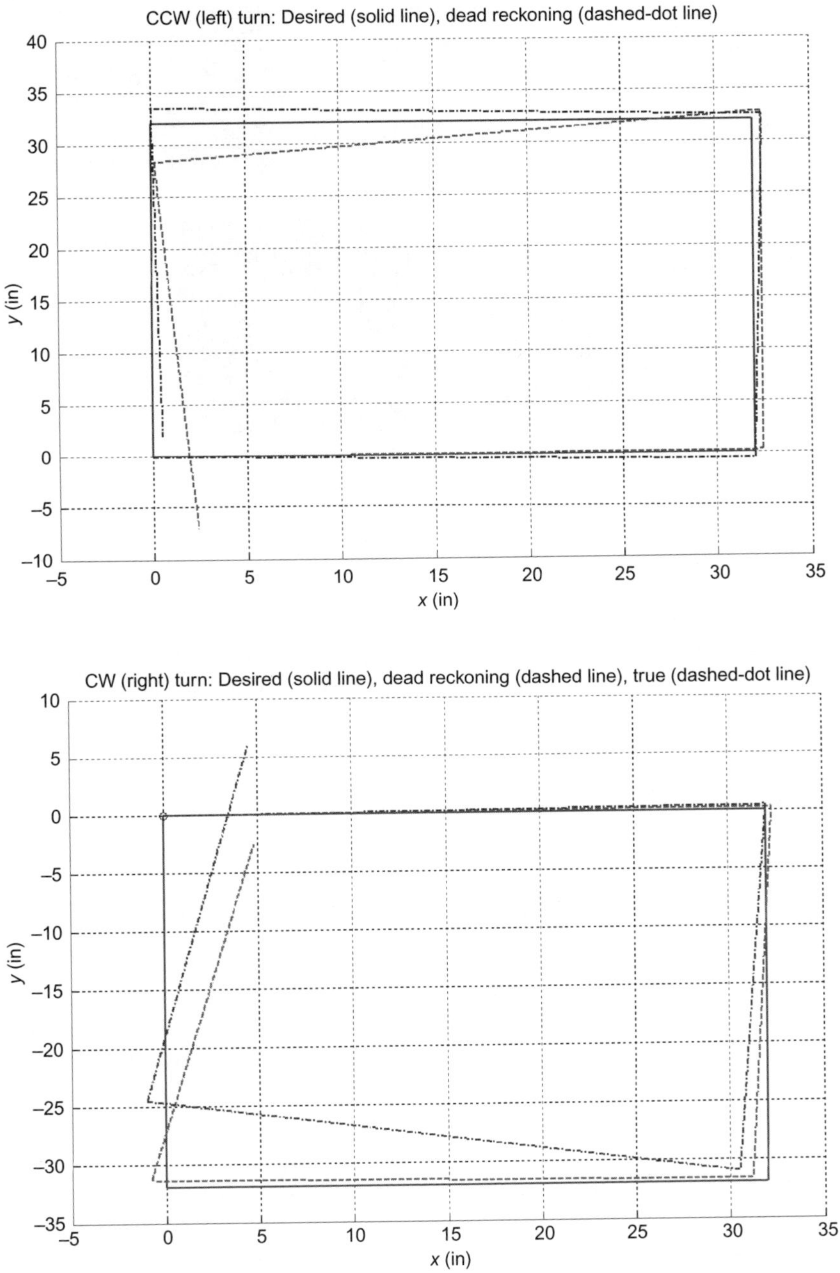


*Figure 2.6* *Results of UMBmark (square path) test for measuring dead-reckoning error. Box of 32 in × 32 in along which robot navigates in counterclockwise (top) and clockwise (bottom) direction*

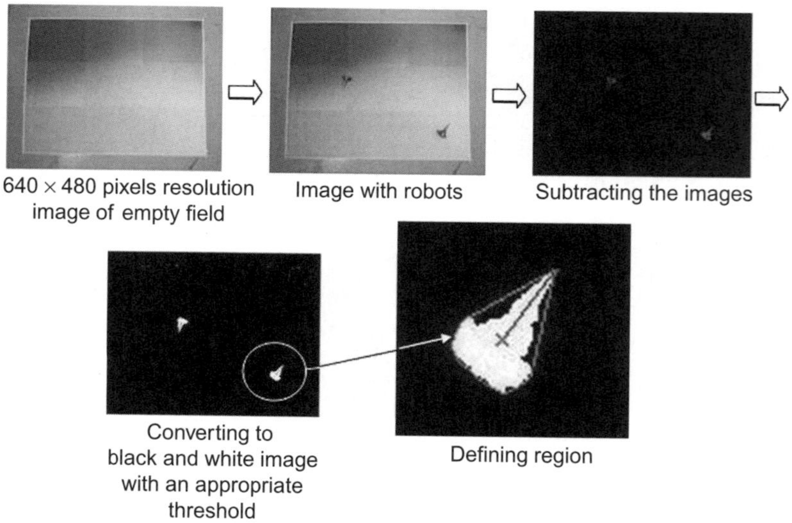


*Figure 2.7 Steps involved in image processing for robot localization*

difference in time of arrival of both messages (RF travels about $10^6$ times faster than ultrasound) the listener can infer its distance from the beacon. If the listener robot can get this distance from three static nodes with known location, it can easily triangulate to find its relative location.

## 2.2 Mobile WSN test bed

### *2.2.1 Objectives of the test bed*

The mobile WSN test bed has been developed to accomplish the following objectives:

- Provide means of experimentally validating resource allocation algorithms, such as the discrete event controller (DEC) discussed in Chapter 6, for mobile resources. The mobile resources could be the MWSN discussed in prior sections, and the tasks could be sensing or sampling environmental fields.
- Provide means of experimentally validating deadlock-free resource allocation algorithms.
- Provide means of experimentally validating deadlock-free resource allocation algorithms in the presence of 'routing' resources – where multiple similar resources exist and for a particular task it does not matter which one of these resources is scheduled.

### *2.2.2 Structure of the test bed*

The mobile wireless sensor network (WSN) test bed at the ARRI, University of Texas at Arlington, consists of mobile sentry robots, unattended ground sensors

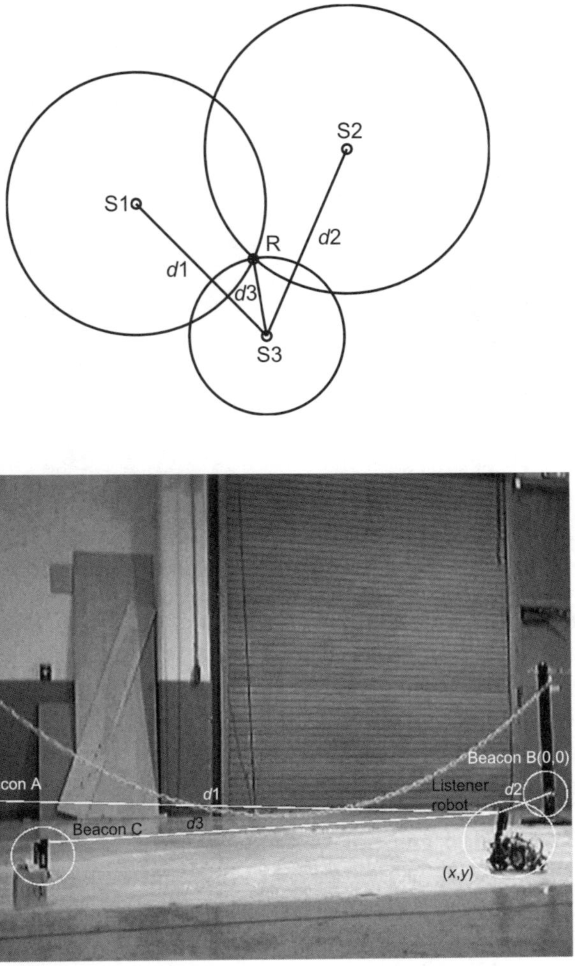


*Figure 2.8 Robot localization from static beacons by triangulation*

(UGS) and a centralized control unit that communicate wirelessly with each other. The control architecture presented in Figure 2.9 has been implemented using LabVIEW. The sensor network is arranged according to a star topology where the base station personal computer (PC) plays the role of the central node. The PC is connected to a transceiver for wireless communication with all the resources (Figure 2.10 (see p. C1)).

### 2.2.2.1 Mobile sentry robots

Two Cybermotion SR2 mobile sentry robots (donated by JC Penney, Inc.) formerly used to patrol a warehouse in Dallas, Texas, are employed as mobile sensing units. They have an extensive sensor suite including ultrasonic intrusion, optical flame detector, dual passive IR, microwave intrusion, smoke, temperature, humidity and

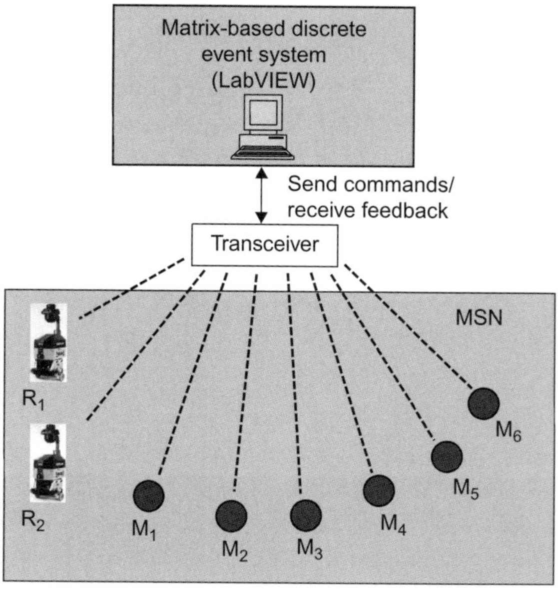


*Figure 2.9 Functional architecture of the ARRI MSN test bed*

light sensors, and gas sensors including oxygen, $NO_x$ and CO. Each robot's task is executed through an *ad hoc* LabVIEW® VI. For the sake of simplicity, manipulation tasks have been implemented just as time delays. However, this does not affect the behaviour of our DEC, which is the focus of this work.

#### 2.2.2.2 Unattended ground sensors

A set of six Berkeley Crossbow UGS has been incorporated into the Secure Area Test bed at the ARRI, University of Texas at Arlington. They can measure various quantities such as light, acceleration, temperature, magnetism and sound. Each mote carries an ATMEL processor running the Berkeley's RTS operating system, TinyOS. To relieve the user from low-level programming, we have developed a LabVIEW® interface to install TinyOS programs on the mote in a completely transparent way for the user. Also, after the motes have been programmed, our LabVIEW interface allows us to change the fly program parameters (such as baud rate and mote identity (ID)) and to perform all the read/write operations from/to the motes.

*Part II*

# Single-robot adaptive sampling

# *Chapter 3*
# Adaptive sampling of parametric fields

This chapter discusses background work related to the deployment of mobile robots for sampling. Section 3.1 presents different sampling strategies. Section 3.2 describes the density estimation by sampling and sensor fusion. Sections 3.3 and 3.4 explain existing approaches for sampling using static and mobile sensor nodes, respectively. Parametric and non-parametric solutions to the sampling problem are also discussed. Section 3.5 presents the existing approaches for reducing the localization error in mobile robots while sampling.

Following this, the chapter focuses on the mathematical formulation of adaptive sampling (AS) problem for parameterized fields, including models, uncertainties and sampling criteria. Section 3.6 discusses sampling strategies such as raster scanning, random sampling, AS and greedy AS (GAS) and provides both a qualitative and a quantitative definition for the AS problem. Section 3.7 presents the extended Kalman filter (EKF) formulation of the AS problem with a single mobile sensor node.

## 3.1 Sampling

The idea of sampling comes from statistics where only a small set of measurements is available, and it is used to estimate the characteristics of a population. Thompson [71] discusses several ways to select few samples from a large data set. When considering sequential estimation of a field distribution, the simplest method is to randomly select samples for every measurement with each item in the population having equal probability for being chosen irrespective of the field. In other words, each sampling point is selected independently from all other sampling points as shown in Figure 3.1a. Another approach is *cluster* or *stratified sampling*, in which the sampling area is classified into homogenous clusters. Such classified strata can be overlapping or non-overlapping. The formation of strata requires prior information about the field of interest. In practice, a combination of these approaches is used, for example, a simple *random sampling* (Figure 3.1a) or *stratified random sampling* (Figure 3.1b). Another sampling approach is *systematic sampling* (*uniform grid-size sampling*), which involves dividing the sampling area into equal-sized grids. Examples of these approaches are *systematic raster scan sampling* (Figure 3.1c) and *systematic random sampling* (Figure 3.1d). In *AS*, sampling locations are selected in real time, based on the information that is gained by past

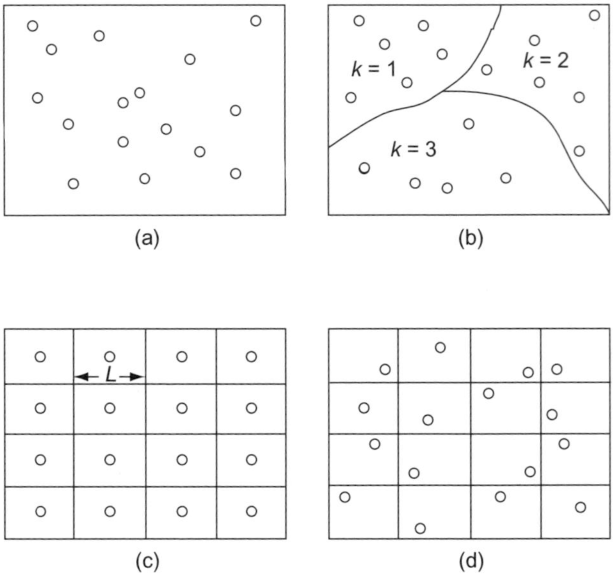


*Figure 3.1 Sampling strategies: (a) simple random sampling, (b) stratified random sampling, (c) systematic raster scan sampling and (d) systematic random sampling [71]*

sampling. Here an AS algorithm for complex parametric fields is presented, and its advantages over random or raster non-AS are highlighted.

## 3.2 Sampling for density estimation

One purpose for sampling a space–time field is density estimation, e.g. the construction of a field estimate of how a certain parameter varies in space and time based on observed data. Data analysis can be performed by assuming either a parametric or a non-parametric distribution. Parametric formulations assume a functional form of the probability density function (PDF), while non-parametric solutions do not assume any particular distribution in advance. In the parametric case, given a density $f(\cdot/\theta)$, such as a two-parameter normal family $N(\mu, \sigma^2)$, where $\theta = (\mu, \sigma^2)$, the emphasis is on obtaining the better estimator $\hat{\theta}$ of $\theta$, and the error criteria can be $MSE\{\hat{\theta}\} = E[\theta - \hat{\theta}]^2$. In the non-parametric case, the emphasis is directly on obtaining a good estimate $\hat{f}(\cdot)$ of the entire density function $f(\cdot)$, and the error criteria can be $MSE\{\hat{f}(\theta)\} = E[f(\theta) - \hat{f}(\theta)]^2$.

There are two types of information for density estimation: *a priori information* and *empirical information*. *A priori* information is the knowledge that was already there before a given observation became available. The combination of prior knowledge and empirical knowledge leads to *a posteriori information*. Parametric description of distributions can be either a physical model of the system or a black or grey box model acquired by clustering the *a priori* data, and then fitting a linear or non-linear model to it. Non-parametric formulations can also involve clustering. The concepts of clustering, parametric approximation and parameter estimation are introduced in this section in the context of using *a priori* data for clustering and system identification to acquire a parametric description, and therefore improve the parameter estimates when empirical knowledge is available.

## *3.2.1 Clustering*

Clustering is the process of organizing a set of data into groups in such a way that observations within a group are more similar to each other than they are to observations belonging to a different cluster. Clustering can be done either by *supervised* or by *unsupervised learning* of the data. Examples of clustering are *hierarchical clustering, k-means clustering, mixture of Gaussians clustering*, etc., the details of which can be found in the literature [72]. Two examples of *k-means clustering* and *fuzzy c-means clustering* (FCM) are relevant and are further covered in this section.

### 3.2.1.1 *k*-Means clustering

*k*-Means clustering is one of the simplest unsupervised learning algorithms for clustering discussed in detail in the literature [73]. The objective is to classify a given set of data into $k$ clusters where each cluster has a centroid $c$. There are three ways to select centroid locations: randomly, at fixed distances or at locations where significant data is present. Once the centroid locations are selected, the next step is to assign each of the available data to the nearest centroid. When all the data is assigned, the locations of centroids are recalculated. The procedure is repeated until the centroids no longer move. The objective function that is minimized for recalculating the centroid is

$$J = \sum_{i=1}^{N} \sum_{j=1}^{K} ||x_i - c_j||^2 \tag{3.1}$$

where $N$ is the number of Gaussian centres, $K$ is the number of clusters and $c_j$ is the centroid of the cluster $j$.

### 3.2.1.2 Fuzzy *c*-means clustering

The fuzzy clustering approach was introduced by Bezdek [74]. When compared to *k*-means clustering, fuzzy clustering allows one piece of data to belong to two or more clusters. Each Gaussian centre belongs to a cluster to some degree that is defined by membership grade by a value $u$ between 0 and 1. The algorithm involves the minimization of the cost function:

$$J_m = \sum_{i=1}^{N} \sum_{j=1}^{C} u_{ij^m} ||x_i - c_j||^2, \quad 1 \leq m < \infty \tag{3.2}$$

where $N$ is the number of Gaussian centres, $C$ is the number of clusters, $u_{ij}$ is the degree of membership of centre $x_i$ in cluster $j$, $c_j$ is the centroid of the cluster $j$ and

$m$ is a real number greater than 1. The member function and cluster centroids are iteratively optimized by the following equations:

$$u_{ij} = \frac{1}{\sum_{k=1}^{C} \left( ||x_i - c_j|| / ||x_i - c_k|| \right)^{2/(m-1)}}, \qquad c_j = \frac{\sum_{i=1}^{N} u_{ij}^m x_i}{\sum_{i=1}^{N} u_{ij}^m} \tag{3.3}$$

This iteration stops when $max_{ij}\{|u_{ij}^{(k+1)} - u_{ij}^{(k)}|\} < \varepsilon$, where $\varepsilon$ is the termination criterion and $k$ is the number of iterations.

### 3.2.2 Parametric approximation

Parametric approximation is a curve-fitting problem, the objective of which is to use the *a priori* information about a non-parametric field to approximate it with a parametric field. Methods for parametric approximation are well described by Nelles [73], and some of the ones used here are covered below.

#### 3.2.2.1 Linear and polynomial approximations

The simplest of parametric approximation is linear for the fields whose non-linear characteristic is weak. For example,

$$\hat{F} = a_0 + a_1 u_1 + a_2 u_2 + \cdots + a_p u_p = \sum_{i=0}^{p} a_i u_i, \quad \text{with} \quad u_0 = 1 \tag{3.4}$$

where $a_i$ are the slope values in the direction of $u_i$ and $a_0$ is the ordinate value at $u = 0$. Equation (3.4) can be consolidated in vector form through a parameter vector:

$$\hat{A} = [a_0 \ a_1 \ \cdots \ a_p]^{\mathrm{T}} \tag{3.5}$$

For non-linear fields, a polynomial approximation can be considered, which is a straightforward extension of a linear model. The higher the degree of the polynomial, the more flexible the model is. For instance, a three-dimensional polynomial of degree 2 is given by

$$\hat{F} = a_0 + a_1 u_1 + a_2 u_2 + a_3 u_3 + a_4 u_1^2 + a_5 u_1 u_2 + a_6 u_1 u_3 + a_7 u_2^2 + a_8 u_2 u_3 + a_9 u_3^2 \tag{3.6}$$

Equations (3.4) and (3.5) represent static fields for which the values of these parameters stay the same. But time-varying fields can also be modelled through time-dependent parameters.

#### 3.2.2.2 Neural network approximation

A general formulation of parametric approximation is an artificial neural network. Instead of assuming a linear or polynomial basis, a neural network assumes a general non-linear basis function $\theta_i^{(nl)}$, and is organized in multiple layers. The most common neural network architectures are MLP (multi-layer perceptrons) and RBF (radial basis function) networks. Both MLPs and RBFs are universal approximators, e.g. they can approximate any smooth function [73]. The universal approximation capability is an important property that will be used here to approximate complex fields. Although MLP has its own advantages in terms of accuracy, its hidden layers are either *log* or *tanh*, which are global activation functions. It may be good to approximate a complex field with parameters of *log* function layer, but it is difficult to distribute computations with global activation functions because the information is not locally contained. In contrast, RBFs' activation is local, which is advantageous in sequential robotic sampling, because samples have an effect on neighbouring Gaussians only. This is further elaborated in Chapter 5 where complex fields are considered. In short, in RBF networks, the parameters of the hidden layers have a more appropriate sampling interpretation than MLP.

An RBF network consists of three types of parameters:

- Output layer weights, $a_i$, are linear parameters. They determine the height of the basis functions and the offset value.
- Centres, $c_{ij}$, are non-linear parameters of the hidden layer neurons. They determine the positions of the basis functions.
- Standard deviations, $\sigma_i$, are also non-linear parameters of the hidden layer neurons. They determine the width of the basis functions.

RBFs exhibit good approximation properties of their own. Any continuous function can be approximated with RBF neural network according to universal approximation theorem.

Let $G\colon \Re^{m_0} \rightarrow \Re$ be an integrable bounded function such that $G$ is continuous and $\int_{\Re^{m_0}} G(x)dx \neq 0$. Let $\Im_G$ denote the family of RBF networks consisting of functions $F\colon \Re^{m_0} \rightarrow \Re$ represented by

$$F(x) = \sum_{i=1}^{N} a_i G\left(\frac{x - c_i}{\sigma_i}\right) \tag{3.7}$$

where $\sigma_i > 0$, $a_i \in \Re$ and $c_i \in \Re^{m_0}$ for $i = 1, 2, \ldots, N$. The universal approximation theorem for RBF networks can then be stated as

*For any continuous input–output mapping function f(x), there is an RBF network with a set of centres $c_i$ and width $\sigma_i$, where $i = 1, 2, \ldots, N$ such that the input–output mapping function F(x) realized by the RBF network is close to f(x) in the $L_p$-norm, $p \in [1, \infty]$* [75].

For training an RBF neural network, different strategies exist. In the simplest case, the hidden layer of the network is not optimized, which means that there is no optimal learning strategy for $c_i$ and $\sigma_i$. But there are several strategies for placing the centre locations before learning of other parameters begins. One such strategy is the random centre placement, which is considered a reasonable initialization method for some of the advanced learning schemes [73]. A straightforward improvement of random selection of centres is the application of clustering techniques discussed in the previous section. The greatest advantage of using this approach is that the centres are selected according to the training data distribution in input space. Therefore, more RBFs are placed in regions where data is dense, and few RBFs are placed in regions where data is sparse.

The universal approximation theorem is valid even for a constant $\sigma$ for all neurons, but there are heuristic methods to calculate standard deviations for training the RBF network. The most common is the *k-nearest neighbour* rule, which assigns each RBF a standard deviation based on average distance between its centre and the centre of the $k$-nearest neighbour RBFs. Finally, the weights of the output layer can be computed using least squares. Both clustering and least squares (LS) algorithms are very fast, even for relatively complex fields.

### *3.2.3 Parameter estimation*

Parameter estimation problems involve finding the best value for parameters of a process based on noisy measurements. Least squares estimation is the most common criterion used for batch as well as sequential estimation problems. For recursive estimation problems, an initial estimate and uncertainty (error covariance) is assumed. As noisy measurements become available sequentially, the parameter estimates, as well as the estimation error covariance, are updated. The Kalman filter is a fundamental estimation tool, which uses the dynamic description of the system along with recursive least squares estimation for processing measurements. Robot location states and field parameter estimation using Kalman filters are covered in great detail in the subsequent sections.

## 3.3 Sampling using static wireless sensor network

Wireless networks of spatially distributed sensors provide an attractive approach to spatially monitor the environment. The obvious advantage of using static nodes compared to mobile robots is that there are no localization errors due to navigation. However, using higher densities of sensors provides more measurements, higher resolution and better accuracy, at the expense of more communication and processing. Several deployment schemes for static nodes have been recently proposed to reduce communication and energy consumption. For the simplest scenario of field estimation, sensors are deployed at the centres of uniformly spaced square grids [76]. Based on the initial information about the field to be mapped, the sensors could be deployed more densely in the high-variance areas and far apart in the areas with small variation [11]. Usually the objective is to minimize the energy consumption and communications by activating only a fraction of sensors for measurements and communication while maintaining high accuracy. However, this kind of approach cannot be used to monitor wide areas. Also, a sparse density of placement of static nodes reduces the accuracy of the estimate.

Other related problems for static sensor network are *coverage* and *connectivity* [77]. The coverage problem is defined as the maximization of the total area covered by the sensors, whereas connectivity is the requirement that every sensor should be able to communicate with every other sensor. There are many applications of coverage and connectivity such as target tracking and search and rescue.

Static nodes are also widely used in applications involving navigation and controlling of mobile robots, environmental monitoring, disaster and rescue work, healthcare, etc. When used for tracking mobile targets, static sensors act as anchor nodes with known position and help the mobile robots localize.

## 3.4 Sampling using robotic sensor deployment

Mobile robots are being increasingly used on land, underwater and in air, as sensor-carrying agents to perform sampling missions, such as searching for harmful biological and chemical agents, search and rescue in disaster areas, and environmental mapping and monitoring. A multitude of research groups have published results on sampling using mobile robots for chemical plume source localization, mine detection, ocean sampling, forest mapping, etc. [10,11,14–16,78–83].

One of the problems most commonly discussed in mobile wireless sensor networks (WSNs) is team objectives for exploration and mapping in an unknown environment. Sensor deployment for sampling can be considered a variant of the exploration and mapping problem. The aim in mapping and exploration is to build a global map of the environment by sequentially visiting each location with one or more robots, whereas in information-aware sampling the aim is to choose the best locations that minimize the uncertainty of the field estimate. In exploration for building a global map, it is impractical to have the robots take measurements at all locations.

When using mobile robots for sampling missions, accurate localization is needed because the uncertainty in localization affects the estimate of the spatio-temporal field. Even though mobility introduces additional degree of complexity for navigation and localization, it allows the repositioning of onboard sensors, thus expanding the coverage and operational lifetime of the sensor network. For instance, Howard *et al.* [77] discuss the mobile robotic sensor deployment using potential fields for maximization of area covered by the network. In addition to coverage, the problem of distributed field variable estimation by inverse modelling of the measurements has been addressed for oceanography and meteorology [84]. Popa *et al.* [85] propose using a potential field framework to control the behaviour of the mobile sensor nodes by combining classical robotic team concepts (obstacle avoidance, goal attainment, flight formation, environment mapping and coverage) with traditional sensor network concepts (node energy minimization, optimal data rate and congestion control, routing in *ad hoc* networks).

In the end, the performance of any deployment scheme for sampling will also depend on the type of field distribution and its representation.

### *3.4.1 Parametric field representation*

Parametric representation is a model-based method in which parameters of a process model are estimated using techniques such as the Kalman filter, least squares estimation or other optimization schemes.

Christopoulos and Roumeliotis [10] present an approach for estimating the parameters of diffusion equation that describes the propagation of an instantaneously released gas. This is accomplished by first building a model considering factors such as the diffusion phenomena of the chemical, wind direction and speed, turbulence, etc. The selection of the set of locations where chemical concentration measurements should be recorded is performed in real time with the objective of maximizing the accuracy of the parameter estimates and reducing the time to convergence of this estimation problem. The selection of sequence of locations where the measurements are recorded is based on the minimization of the uncertainty of the estimated parameter that is represented by the trace of the error covariance of parameter estimates.

Cannell and Stilwell [86] present two approaches for AS of underwater processes using autonomous underwater vehicles (AUVs). The first one assumes a parametric model, whereas the second one uses an information-theoretic approach. For the parametric approach, two methods for estimating the unknown parameters of the process model are considered. The first method is related to mapping algorithms where the unknown parameters in an assumed measurement model are states of a Kalman filter. In the second method, non-linear least squares are used to estimate the parameters.

### *3.4.2 Non-parametric field representation*

A number of strategies for non-parametric AS can also be found in the literature. A solution for non-parametric ocean sampling based on classification of the sampling area is proposed in Reference 11. The area is divided into non-uniform-sized grids, and the size of the grid is selected on the basis of the local variance of the distribution. Hence, the areas with high variance have small-sized grids and the areas with small variance have large-sized grids, which is important for thoroughly sampling the areas with more variation and just taking a few samples in more uniform areas.

In some cases, information-theoretic approaches are utilized, which process data via sequential classification and tracking phases. During the classification phase, measurements are collected in the assumed ambient flow, and an empirical distribution is formed. This distribution represents the measurements taken outside the process. As new measurements are taken, updated empirical distributions are computed and compared to the assumed ambient distribution using the *minimum description length (MDL) test*. This test utilizes the *Kullback–Leibler divergence*, which does not require any explicit defined process model. It characterizes the boundary of a closed *divergence*, also known as *relative entropy*, to determine whether two distributions are different in a statistically significant manner. This result is then used as a test for identifying data that is statistically different from the ambient distribution. This information-theoretic approach has been applied to ocean sampling and prediction [13].

Another non-parametric approach is *cognitive* or *rule-based* in order to adaptively select sample locations in real time based on current and past samples to ensure that a field source is localized. This scheme does not involve formation of any distribution, but it is used to find distribution maxima. This non-parametric approach is used very widely in chemical plume tracing on land and in water [78], odour sensing, mine detection [80], etc.

Chemical source localization is usually a gradient climbing problem with disturbances caused by wind, gas diffusion and turbulence from the obstacles and the environment. A lot of research has been done in the area of odour source localization, landmine detection, etc., using both parametric and non-parametric approaches. Non-parametric approaches involve taking appropriate samples in a recursive fashion in order to find the source. In practice, such approaches generally involve three phases: detecting the plume, tracking the chemical towards its source and finally locating the source. Some source localization algorithms are biologically inspired because insects and most other animals are very apt at sensing odours in order to find food, locate mates or communicate. The applications for source localization using robots are in locating harmful biological and chemical agents, and avalanche and earthquake victims, detecting landmines, etc. It would be difficult for humans to find chemical sources because of the presence of poisonous gases and nuclear radiation or any other hazardous conditions.

*Chemotaxis* and *anemotaxis* are the most common techniques used for estimating chemical distributions. Chemotaxis relies on the local gradient of the chemical agent concentration, whereas anemotaxis-based approaches require that the agent move in the upwind direction. The robots used for source localization are equipped with anemometric sensors and gas sensors, and in some cases with vision. Several algorithms such as step-by-step, zigzag and active sampling can be found in the literature [10,15,16,81–83].

Another similar application of non-parametric sampling is mine detection using demining robots. The sensors used are metal detectors, IR detectors and chemical sensors. Literature on mine detection using robots can be found in Reference 80.

## 3.5 Adaptive sampling problem for robots

Mobile robots are being used on land, underwater and in air, as sensor-carrying agents to perform sampling missions, such as searching for harmful biological and chemical agents, search and rescue in disaster areas, and environmental mapping and monitoring. Sampling can be simply defined as *a set of elements drawn from and analysed to estimate the characteristics of a population*. In AS, *strategies continuously adapt in response to real-time environmental measurements without human intervention*. Once such scenario involved the deployment of multiple underwater vehicles for the environmental monitoring of large bodies of water, such as oceans, harbours, lakes, rivers and estuaries [13,78,86]. Predictive models and maps can be created by repeated measurements of physical characteristics, such as water temperature, dissolved oxygen, current strength and direction, or bathymetry. However, because the sampling volume could be large, only limited measurements are available. Intuitively, a deliberate sampling strategy based on models will be more efficient than just a random sampling strategy.

Popa *et al.* [13] described a combined multi-agent AS problem coupling uncertainty in localization as well as in the sensor measurements to achieve effective AS using a solar AUV, considering very simple, linear field models. Localization uncertainties are relevant to the sampling problem since position estimates for underwater robots are often inaccurate due to navigational errors from dead reckoning. Here we expand and extend this parametric approach to include much more complex fields and heuristics with much better sampling performance. Unlike non-parametric sampling methods, such as Reference 11, our approach requires a parametric field description of the sampled field.

Qualitatively, a multi-agent AS problem for spatio-temporal fields can be posed as follows:

*We wish to describe an unknown non-linear spatio-temporal field variable via a parametric approximation $Z = Z(A, X, t)$ depending on an unknown parameter vector A, position vector X and time t. The field is recovered by using N robotic vehicles, sampling the field with localization and sensing uncertainty. We wish to*

- *Decide what sampling locations $X_k^i$, $1 \leq i \leq N$, $1 \leq k \leq n$, should be chosen such that the uncertainty in estimating the unknown parameter vector is minimized.*
- *Additional secondary objectives such as the number of samples, sampling duration, energy, etc., are minimized.*
- *Distribute the computational and communication load, and the sampling space among N robots in an efficient manner.*

## 3.6 Sampling strategies: where to sample

### *3.6.1 Global search AS*

In the absence of model and measurement uncertainties in localization and field parameters, the sampling task would be a deterministic case, and all the sample locations $X_i$ $(1 \leq i \leq N)$ will provide same information about the unknown $X$ and $A$. Hence, there would be no difference in random, row-by-row or AS. But in the presence of uncertainties discussed, the error covariance (uncertainty) of states can be used to decide the next best sampling location. The error covariance of the parameter states is the measure of uncertainty of the parameters. High value of error covariance indicates higher uncertainty, and lower value indicates lower uncertainty.

AS location calculates the anticipated error covariance of all the candidates' location and selects the one that gives minimum value. The minimization in the error covariance of a parameter is achieved by sampling along the extreme (boundary) values of the parameter. A summary of different ways to pick the next sampling location is shown in Figure 3.2.

It will be shown in the next section through numerous simulation and experimental results that AS requires considerably fewer samples than raster scanning or

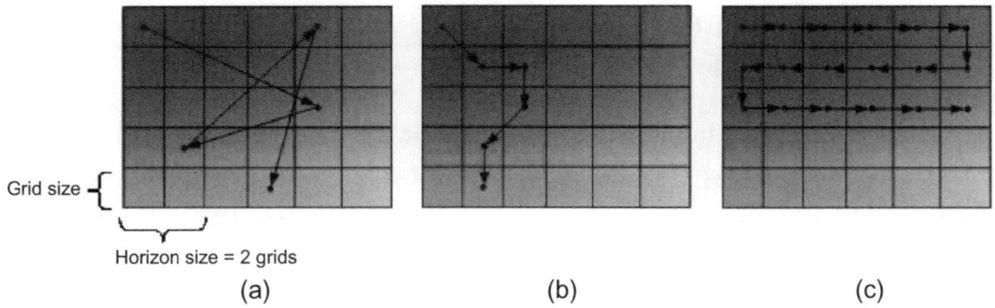


*Figure 3.2 (a) Adaptive sampling with uniform grid size, (b) greedy AS with uniform grid size and uniform step size and (c) raster scanning*

random sampling because AS minimizes the uncertainty in unknown parameters. However, one drawback of using this standard approach of AS is that it is computationally complex because it calculates the anticipated error covariance of the locations in the entire search space. To overcome this drawback, a heuristic GAS is used to indirectly incorporate the time to move for the robot and reduce complexity.

### *3.6.2 Heuristic greedy AS*

If the sampling area is large, the optimal sampling position could be far away from current robot location, and it will not be efficient in minimizing the sampling time. A heuristic greedy approach is adopted instead, where the next sampling location is searched within the vicinity of the currently sampled location. The advantage is that the robot covers less distance that saves time and number of samples. Also, it reduces the complexity of the algorithm because only the anticipated error covariance of sampling neighbouring locations is calculated. The heuristic GAS algorithm helps to answer the following question:

*Given a parametric field variable $Z = Z(A, X)$ depending on an unknown parameter vector A and position vector X, and a robotic vehicle/sensor node navigating the field with localization uncertainty, what sampling locations $X_i$ (where $1 \leq i \leq N$), in the immediate vicinity of the current sampling location, should be chosen such that a trade-off between the uncertainty in estimating the unknown parameter vector and the sampling distance is minimized?*

## 3.7 Basic EKF formulation

To pose the AS problem in a quantitative way, an information measure and uncertainty model needs to be selected. We use the EKF where the state transition and measurement equations during sampling can be described through non-linear mappings (functions *h, m, f* and *g*) as follows:

*Robotic sensor nodes state dynamics*:

$$X^i_{k+1} = X^i_{k+1} + h(X^i_k, U^i_{1k}) + w_k{}^i \tag{3.8}$$

where $X_k{}^i$ and $U_k{}^i$ are the *i*th vehicle state and control input, and the state has noise covariance matrix $E[w_k{}^i(w_k{}^i)^{\mathrm{T}}] = Q^i_{1k}$.

*Field parameter dynamics*:

$$A_{k+1} = A_k + m(A_k, U_{2k}) + \alpha_k \tag{3.9}$$

where $A_k$ is a vector of unknown coefficients describing the field with noise covariance matrix $E[\alpha_k {\alpha_k}^{\mathrm{T}}] = Q_{2k}$ and $U_{2k}$ is the uncontrollable (but measurable) 'field evolution vector'. This parameter is a slow-varying correction factor in the field parameters, assuming that infrequent, low-resolution measurements of the entire field are available.

*Robot position output measurement*:

$$Y_k^i = f(X_k^i) + \xi_k^i \tag{3.10}$$

where the output noise covariance is $E[\xi_k^i (\xi_k^i)^{\mathrm{T}}] = R_{1k}^i$.

*Distributed field variable measurement model*:

$$Z_k^i = g(X_k^i, A_k) + \nu_k^i \tag{3.11}$$

where ${Z_k}^i$ is the field variable with measurement noise covariance $E[\nu_k^i (\nu_k^i)^{\mathrm{T}}] = R_{2k}^i$.

*Robot input measurement*:

$$\tilde{U}_{1k}^i = U_{1k}^i + \beta_k^i \tag{3.12}$$

where $\tilde{U}_{1k}^i$ is the measured control input with measurement noise covariance $E[\beta_k^i (\beta_k^i)^{\mathrm{T}}] = R_{3k}^i$. $U_{1k}^i$ is the output caused by the control input $\tilde{U}_{1k}^i$.

*Field evolution factor measurement*:

$$\tilde{U}_{2k} = U_{2k} + \eta_k \tag{3.13}$$

where $\tilde{U}_{2k}$ is the measured field parameter evolution using high spatial and low rate measurement with measurement noise covariance $E[\eta_k {\eta_k}^{\mathrm{T}}] = R_{4k}$. $U_{2k}$ is the actual change in the parameter value since the last estimate.

For the single mobile agent $i$ sampling the field, the EKF overall state vector is $(X_k^i, A_k)$, and the system and measurement models are thus summarized as

$$\begin{aligned}
\begin{pmatrix} X_{k+1}^i \\ A_{k+1} \end{pmatrix} &= \begin{pmatrix} X_k^i \\ A_k \end{pmatrix} + \begin{pmatrix} h(X_k^i, U_{1k}^i) \\ m(A_k, U_{2k}) \end{pmatrix} + \begin{pmatrix} {w_k}^i \\ \alpha_k \end{pmatrix} \\
{Y_k}^i &= f({X_k}^i) + {\xi_k}^i \\
{Z_k}^i &= g({X_k}^i, A_k) + {\nu_k}^i \\
\tilde{U}_{1k}^i &= U_{1k}^i + {\beta_k}^i \\
\tilde{U}_{2k} &= U_{2k} + \eta_k
\end{aligned} \tag{3.14}$$

Using the standard non-linear, discrete EKF formulation [24], the time and measurement update equations are given by

$$P_{k+1}^{-} = P_k + Q_k, \qquad \begin{pmatrix} \hat{X}_{k+1}^{i-} \\ \hat{A}_{k+1}^{-} \end{pmatrix} = \begin{pmatrix} \hat{X}_k^{i} \\ \hat{A}_k \end{pmatrix} + \begin{pmatrix} h\left(X_k^i, \tilde{U}_{1k}^i\right) \\ m\left(A_k, \tilde{U}_{2k}\right) \end{pmatrix}$$

$$G_k^i = \frac{\partial g_k}{\partial\left(\hat{X}_k^{i-}, \hat{A}_k^{-}\right)}, \qquad F_k^i = \frac{\partial f_k}{\partial \hat{X}_k^{i-}}, \qquad H_k^i = \begin{bmatrix} F_k^i \\ G_k^i \end{bmatrix}$$

$$P_{k+1}^i = \left(\left(P_{k+1}^{i-}\right)^{-1} + H_{k+1}^{i\mathrm{T}}\left(R_k^i\right)^{-1} H_{k+1}^i\right)^{-1} \tag{3.15}$$

$$\begin{pmatrix} \hat{X}_{k+1}^{i} \\ \hat{A}_{k+1} \end{pmatrix} = \begin{pmatrix} \hat{X}_{k+1}^{i-} \\ \hat{A}_{k+1}^{-} \end{pmatrix} + P_{k+1}^i H_{k+1}^{i\mathrm{T}}\left(R_{k+1}^i\right)^{-1}\left[\begin{pmatrix} Y_{k+1}^i \\ Z_{k+1}^i \end{pmatrix} - \begin{pmatrix} f_{k+1}\left(\hat{X}_{k+1}^{i-}\right) \\ g_{k+1}\left(\hat{X}_{k+1}^{i-}, \hat{A}_{k+1}^{-}\right) \end{pmatrix}\right]$$

where the state is the combined position of all the robots and the field parameter vector $[X \;\; A]^{\mathrm{T}}$, $P_k$ is the state covariance, and the process and measurement error covariances are given by

$$Q_k{}^i = \begin{bmatrix} Q_{1k}^i & 0 \\ 0 & Q_{2k} \end{bmatrix}, \quad R_k{}^i = \begin{bmatrix} R_{1k}^i & 0 & 0 & 0 \\ 0 & R_{2k}^i & 0 & 0 \\ 0 & 0 & R_{3k}^i & 0 \\ 0 & 0 & 0 & R_{4k} \end{bmatrix} \tag{3.16}$$

By using a norm of the state covariance as information measure, the AS algorithm moves the vehicle $i$ from $X_k{}^i$ to $X_{k+1}^i$ such that the following $p$-norm of the covariance matrix is minimized over the search space $\Theta$:

$$m(X_k) = \|P_k^i\|_p, \quad m(X_{k+1}^i) \leq m(X_k^i), \quad \forall X^i \in \Theta \tag{3.17}$$

The above formulation applies to both single and multiple robotic sensor agents on a spatio-temporal field with location and field parameter measurements available at the same rate. In past work, several cases were considered by Sreenath and co-workers [50,87] to estimate the parameters of stationary field using least squares and the Kalman filter with single robot, as detailed below.

### *3.7.1 Least squares estimation for linear-in-parameters field*

If the dynamics of the robot is ignored, and if we assume that the parametric field measurement $Z$ depends linearly on the unknown parameter vector $A$, but non-linearly on position vector $X$, then a closed-form solution on the unknown parameters can be obtained, subject to the condition that there is no process and position measurement noise. After $n$ scalar measurements $Z_1, \ldots, Z_n$ are taken at location vectors $X_1, \ldots, X_n$, the field equations can be used to find the least squares solution:

$$
\begin{aligned}
Z_1 &= a_0 + a_1 g_1(X_1) + \cdots + a_m g_m(X_1) \\
Z_2 &= a_0 + a_1 g_1(X_2) + \cdots + a_m g_m(X_2) \\
&\vdots \\
Z_n &= a_0 + a_1 g_1(X_n) + \cdots + a_m g_m(X_n)
\end{aligned}
\tag{3.18}
$$

where $A = [a_0 \; a_1 \; \cdots \; a_m]^{\mathrm{T}}$ is the parameter vector and $g_1, g_2, \ldots, g_m$ are non-linear basis functions of the field. Equation (3.18) can be written in matrix format as

$$
\begin{bmatrix} Z_1 \\ \vdots \\ Z_n \end{bmatrix} = \begin{bmatrix} 1 & 1 & \cdots & 1 \\ g_1(X_1) & g_2(X_1) & \cdots & g_m(X_1) \\ \vdots & \vdots & \ddots & \vdots \\ g_1(X_n) & g_1(X_n) & \cdots & g_m(X_n) \end{bmatrix} \begin{bmatrix} a_1 \\ \vdots \\ a_m \end{bmatrix} = M_n A_n \tag{3.19}
$$

where $M_n$ is an $n \times m$ matrix. The least square solution of (3.19) is simply

$$
\hat{A}_n = (1 \;\; g_i(X_j))^{+}_{i \leq m, j \leq n} \begin{pmatrix} Z_1 \\ \vdots \\ Z_n \end{pmatrix} = {M_n}^{+} \begin{pmatrix} Z_1 \\ \vdots \\ Z_n \end{pmatrix} \tag{3.20}
$$

where ${M_n}^{+}$ is the Moore–Penrose inverse of $M_n$. Moreover, the pseudo-inverse has a closed-form solution given by

$$
{M_n}^{+} = ({M_n}^{\mathrm{T}} M_n)^{-1} {M_n}^{\mathrm{T}} \tag{3.21}
$$

leading to a parameter estimate closed form, as follows:

$$
\hat{A}_n = \left( \sum_{j=1}^{n} (1 \; \cdots \; g_i(X_j) \; \cdots \; g_m(X_j)) \begin{pmatrix} 1 \\ \vdots \\ g_i(X_j) \\ \vdots \\ g_m(X_j) \end{pmatrix} \right)^{-1} \sum_{j=1}^{n} Z_j \begin{pmatrix} 1 \\ \vdots \\ g_i(X_j) \\ \vdots \\ g_m(X_j) \end{pmatrix} \tag{3.22}
$$

Assuming that the robot locations are known without uncertainty, the covariance matrix of $\hat{A}_n$ can be related directly to the constant measurement uncertainty as

$$
\mathrm{var}(\hat{A}_n) = \mathrm{var}(Z_i)({M_n}^{\mathrm{T}} M_n)^{-1} \tag{3.23}
$$

An AS algorithm should then move the robot from $X_n$ to $X_{n+1}$ such that the $p$-norm of the covariance matrix is minimized over the search space $\Theta$:

$$m(X) = \left\| \begin{pmatrix} & 1 \\ & \vdots \\ M_n^{\mathrm{T}} & g_i(X) \\ & \vdots \\ & g_m(X) \end{pmatrix} \begin{pmatrix} & M_n & & & \\ 1 & \cdots & g_i(X) & \cdots & g_m(X) \end{pmatrix} \right\|_p \tag{3.24}$$

or

$$m(X_n) = \|P_n\|_p, \quad m(X_{n+1}) \leq m(X_n), \quad \forall X \in \Theta \tag{3.25}$$

### 3.7.2 *Kalman filter estimation for linear-in-parameters field with no uncertainty in localization*

Like in the previous case, if the parametric form of a measurement field is known, as might be the case, for example, with a bottom profile or systematic variations in temperature or salinity, the field estimation can be integrated with localization in order to improve the estimation process. The assumption that the field distribution is linear in its parameters allows computing a closed-form solution for the information measure used by the sampling algorithm, as we saw in the previous section. The same result can be recovered using the Kalman filter. After taking $n$ measurements at location $X_i$, the field model depends linearly on $m$-dimensional parameter coefficients $A = (a_j)_{1 \leq j \leq m}$ via position-dependent functions as in (3.19) and (3.20).

If the robot dynamics and localization uncertainty are ignored, the state is simply the parameter vector, and KF equations can be written simply as

$$\begin{aligned} A_{k+1} &= A_k + \alpha_k \\ Z_k &= G_k A_k + \nu_k \end{aligned} \tag{3.26}$$

where $E[\alpha_k \alpha_k^T] = Q$, with $Q=0$, $G_k = (1 \;\cdots\; g_i(X_k) \;\cdots\; g_m(X_k))$ and $E[\nu_k \nu_k^T] = R$.

The error covariance matrix from the KF is used to decide where to sample next. Because of the simple state update equation, the KF equation can be reduced to

$$\begin{aligned} & A_o \sim (\bar{A}_o, P_{A_o}), \qquad P_o = P_{A_0} \\ & P_{k+1}^{-1} = P_k^{-1} + G_{k+1}^{\mathrm{T}} R^{-1} G_{k+1} \\ & \hat{A}_{k+1} = \hat{A}_k + P_{k+1} G_{k+1}^{\mathrm{T}} R^{-1}(Z_{k+1} - G_{k+1}\hat{A}_{k+1}) \end{aligned} \tag{3.27}$$

The error covariance is similar to the least squares solution in (3.24) and (3.25) and can be directly calculated by the following formula:

$$P_k = \left( P_0^{-1} + \sum_{j=1}^{k} G_j^{\mathrm{T}} R^{-1} G_j \right)^{-1} \tag{3.28}$$

AS algorithm should then move the robot from $X_n$ to $X_{n+1}$ such that the $p$-norm of the covariance matrix is minimized.

It is apparent that a single scalar value cannot capture all the aspects of the error covariance matrix. For cases involving combined uncertainty of several parameters, information measures that can be considered include the 2-norm, inf-norm, trace or log-determinant of the error covariance matrix, as discussed in Reference 88. The 2-norm of error covariance is a popular choice; however, most of these measures lead to parameter convergence. The *log-det*(*P*) measure can take negative values. *Inf-norm*(*P*) minimizes the uncertainty of the parameter that has the maximum uncertainty, and, as a result, oscillations can be seen before the estimates converge to final values. These oscillations can also cause the EKF filter to diverge due to operating in non-linear regions. *Trace*(*P*), although it shows convergence, might not be appropriate because it does not consider the cross-correlation terms of $P$.

We conclude that the choice of covariance matrix norm will affect the performance of the sampling algorithm. In the absence of any prior information about the field, the 2-norm is an appropriate choice.

If the sampling area is large, the optimal sampling position could be far away from the current robot location, and it will not be efficient in minimizing the sampling time. A heuristic greedy approach (called GAS here) is adopted instead, where the next sampling location is searched within the vicinity of the currently sampled location. The advantage is that the robot covers less distance that saves time and number of samples. Also, it reduces the complexity of the algorithm because only the anticipated error covariance of sampling neighbouring locations is calculated.

#### 3.7.2.1 Simulation results

As an example, consider a stationary linear field, with two spatial variables $(x, y)$ and three unknown parameters to be estimated. The field model is simply defined as

$$Z = a + bx + cy \tag{3.29}$$

For simplicity, assume that we wish to recover the parameters by sampling with a robot with no localization uncertainty. The process, measurement models and update equations are given as follows:

- System model:

$$A_{k+1} = A_k = [a \quad b \quad c]^{\mathrm{T}} \tag{3.30}$$

- Measurement model:

$$\tilde{z}_k = H_k A_k + v_k = [1 \quad x \quad y] A_k + v_k \tag{3.31}$$

- Model and measurement uncertainties: $Q = 0$, $v_k \sim N(0, R)$ and $x_0 \sim N(\bar{x}_0, P_0)$.
- Effect of system dynamics:

$$\hat{A}^{-}_{k+1} = \hat{A}_k, \quad P^{-}_{k+1} = P_k \tag{3.32}$$

- Effect of measurement:

$$\begin{aligned} K_{k+1} &= P^{-}_{k+1} H^{\mathrm{T}}_{k+1} (H_{k+1} P^{-}_{k+1} H^{\mathrm{T}}_{k+1} + R_{k+1})^{-1} \\ P_{k+1} &= (I_{12} - K_{k+1} H_{k+1}) P^{-}_{k+1} \\ \hat{A}_{k+1} &= \hat{A}^{-}_{k+1} + K_{k+1} \{\tilde{z}_{k+1} - H_{k+1} \hat{A}^{-}_{k+1}\} \end{aligned} \tag{3.33}$$

We simulate the sampling process in MATLAB assuming the following numerical values:

- Actual parameters: $[a \quad b \quad c] = [10 \quad 0.2 \quad 0.7]$.
- Initial parameter estimates: $[1 \quad 1 \quad 1]$.
- Parameters to estimate: $A = [a \quad b \quad c]$.
- Initial error covariances: $P = \begin{bmatrix} 10 & 0 & 0 \\ 0 & 10 & 0 \\ 0 & 0 & 10 \end{bmatrix}$.
- Model error covariance: $Q = 0$.
- Measurement error covariance: $R = 1$.
- Convergence criteria: $\|A - \hat{A}_{k+1}\| < 0.7$.

The sampling area is divided into grid sizes of five and ten units, and we assume that only the centre of the grid is sampled. For the GAS heuristic, we pick a horizon size equal to 2 to limit the search for the next appropriate sampling location, as depicted in Figure 3.3.

In this very simple case, there are three parameters to estimate, and the AS algorithm looks for the next sampling location that minimizes the 2-norm of the error covariance matrix of these three parameters. If we wish to focus on minimizing the uncertainty of a specific parameter, instead of the 2-norm aggregate, the next sampling location would be the point that minimizes the error covariance of that particular parameter only. In our simulation, a grid size of 5 is chosen, and therefore a maximum horizon size of 19 will cover the entire field. Simulation results are plotted in Figure 3.4 for a global search AS. The plots are organized to include a snapshot of the first few samples (top left), many samples (top right), the plots of error covariance, the parameter estimate and values of error in parameter estimates with respect to number of samples (bottom left), and the plots of error in parameter estimates and error covariances with respect to the distance travelled (bottom right).

Figure 3.4 shows the results when the information measure of the algorithm is the error covariance of *a* estimates only. We observe that the maximum information

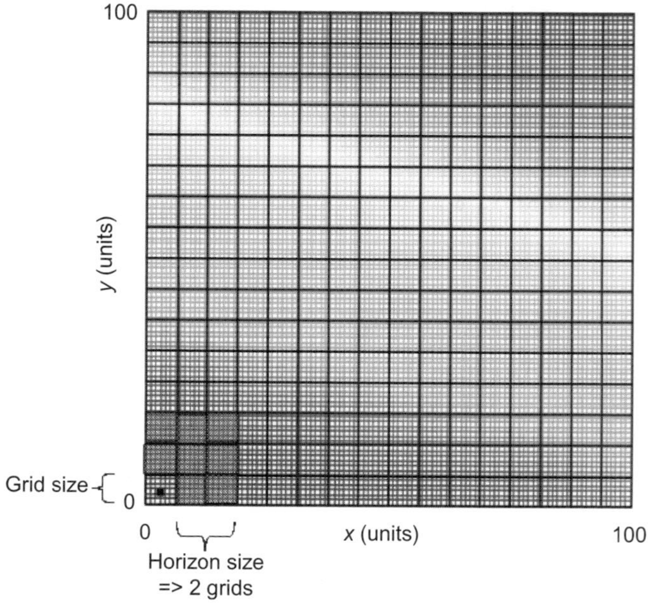


*Figure 3.3 Grid size and horizon size are shown for sampling a linear parametric field*

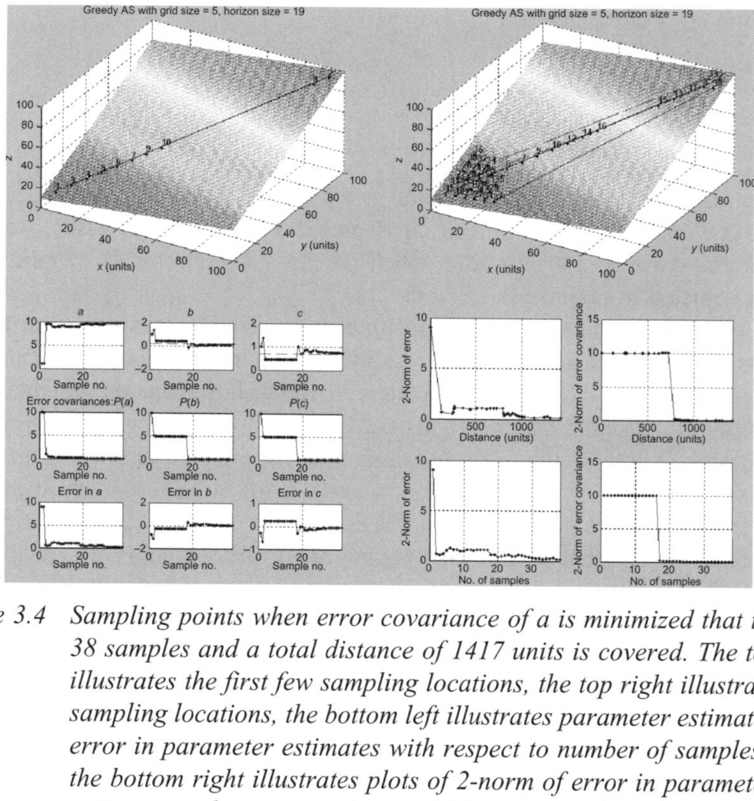


*Figure 3.4 Sampling points when error covariance of a is minimized that takes 38 samples and a total distance of 1417 units is covered. The top left illustrates the first few sampling locations, the top right illustrates all sampling locations, the bottom left illustrates parameter estimates and error in parameter estimates with respect to number of samples, and the bottom right illustrates plots of 2-norm of error in parameter estimates and error covariances with respect to distance travelled, and number of samples*

about parameter $a$ is achieved by first sampling the extreme $(x, y)$ locations. The plots show that parameter $a$ converges faster, i.e. within five samples in this case. Other results that are not presented here indicate that the maximum information about parameter $b$ is achieved by sampling along $x$-axis, first starting at the extreme locations. Similarly, maximum information about parameter $c$ is achieved by sampling along $y$-axis, first starting at the extreme locations. In the last case shown in Figure 3.5, the 2-norm of error covariance of all parameter estimates is used as information measure, and the AS algorithm produces the combined effect to select appropriate locations. In all cases, we also computed the combined distance travelled by the sampling agent, which is displayed in the captions and summarized in Table 3.1.

An interesting observation emerges from examining Table 3.1 – namely that minimizing the error covariance of parameters $b$ and $c$ requires more distance and number of samples to achieve the same convergence threshold. This is due to the fact that sampling along extreme $x$- and $y$-axes by minimizing uncertainty in $b$ and $c$ does not provide enough information about rest of the parameters. This can also be observed in cases where minimization of uncertainty in $b$ causes $a$ and $c$ to converge very slowly compared with the case in Figure 3.5, where minimization of uncertainty in $a$ causes $b$ and $c$ to converge at a significantly faster rate within five samples.

Next, we run the AS algorithm with different horizon and grid sizes and show the comparison in Table 3.2. In Figure 3.6, it is apparent that for raster scan (RS), increasing the grid size reduces the number of samples required to meet the convergence criterion. The number of samples per row is smaller, and hence more rows

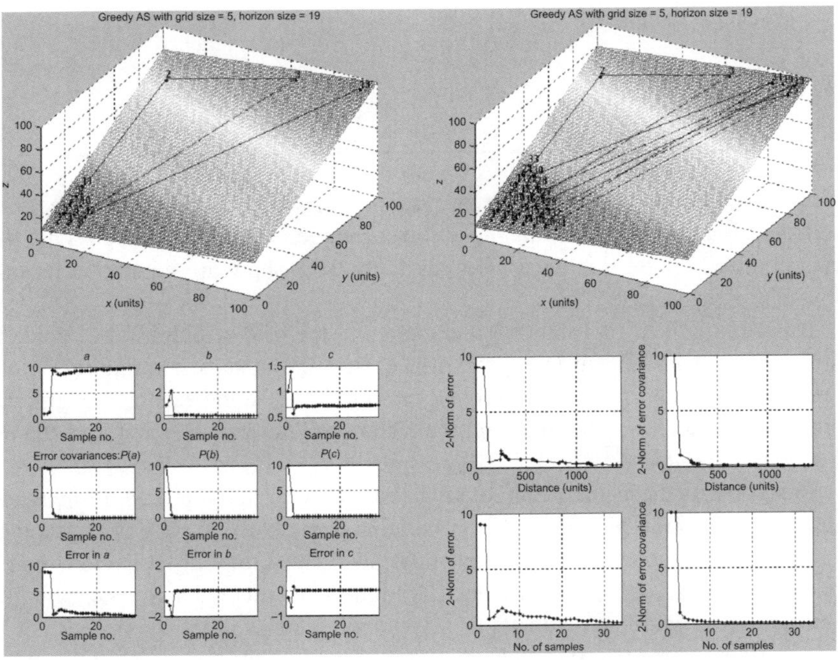


*Figure 3.5 Sampling points when 2-norm of error covariance matrix P is minimized that takes 34 samples and a total distance of 1445 units is covered*

*Table 3.1 Comparison of GAS and AS for number of samples and distance when error covariances of different parameters are minimized*

| | **Minimizing error covariance** | **No. of samples** | **Distance** |
|---|---|---|---|
| Adaptive sampling (grid size = 5, horizon size = 19) | $P(a)$ | 38 | 1417 |
| | $P(b)$ | 68 | 5458 |
| | $P(c)$ | 68 | 5458 |
| | $\|\|P\|\|$ | 34 | 1445 |
| Greedy adaptive sampling (grid size = 5, horizon size = 1) | $P(a)$ | 64 | 412 |
| | $P(b)$ | 128 | 720 |
| | $P(c)$ | 128 | 720 |
| | $\|\|P\|\|$ | 64 | 396 |

*Table 3.2 Comparison of RS, GAS and AS for number of samples and distance when different grid sizes and horizon sizes are considered for sampling a linear field*

| | **Grid size** | **Horizon size** | **No. of samples** | **Distance** |
|---|---|---|---|---|
| Raster scan | 5 | – | 118 | 560 |
| | 10 | – | 78 | 770 |
| Adaptive sampling | 5 | 19 | 27 | 1124 |
| | 10 | 9 | 21 | 1205 |
| Greedy adaptive sampling | 5 | 1 | 62 | 386 |
| | 10 | 1 | 28 | 336 |

are scanned by covering more area on *x*-axis. A larger grid size of ten gives better results because it covers more area with fewer samples, while the field does not have local variance because it is linear. Figures 3.7 and 3.8 show the simulation for other grid and horizon sizes.

It is summarized in Table 3.2 that changing the grid size does not reduce the number of samples and distance for AS because the next best sampling point is searched in the entire area. Moreover, in some cases, increasing the grid size increases the number of samples and the distance to convergence because the data set becomes smaller.

In summary, these simple simulation results reveal that the field discretization grid, the information measure and the horizon size greatly impact the route taken by the robot to accomplish sampling missions. The sampling behaviour can be intuitively understood and shows that GAS requires a much smaller sampling distance than RS, while AS offers the least number of samples at the cost of the greatest travel distance. This observation is still valid for all simulation and experimental results presented here.

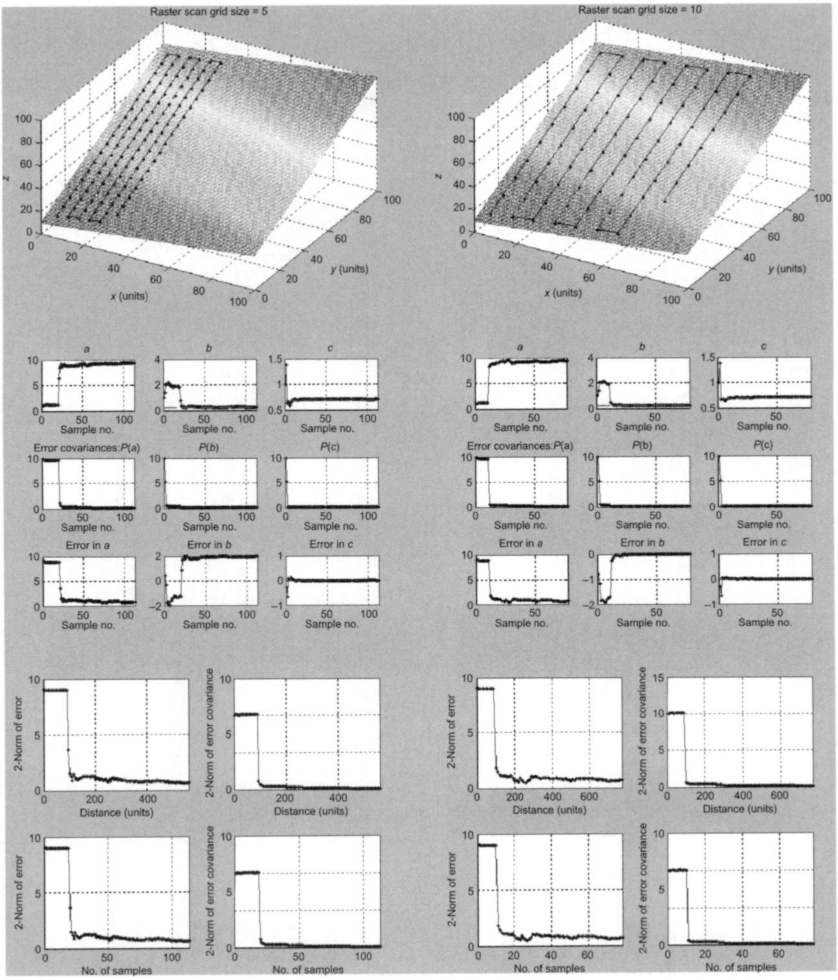


*Figure 3.6* *Simulation results for raster scan when grid sizes of five and ten units are considered*

### 3.7.2.2 Experimental results

Here we present the experimental validation when uncertainty in camera measurement is ignored. In this case, we are recovering a linear colour field with nine parameters; therefore, $X_1 = x$, $X_2 = y$, $g_1(X_1) = x$ and $g_2(X_2) = y$. The base station runs the AS algorithm and commands the mobile sensor to sample at a particular location. The mobile sensor navigates to that location by dead reckoning and thus will end up at a location different than the commanded one. The process and measurement model and update equations of the sampling algorithm are as follows:

- System model:

$$A_{k+1} = A_k = [r_0 \quad r_1 \quad r_2 \quad g_0 \quad g_1 \quad g_2 \quad b_0 \quad b_1 \quad b_2]^{\mathrm{T}} \tag{3.34}$$

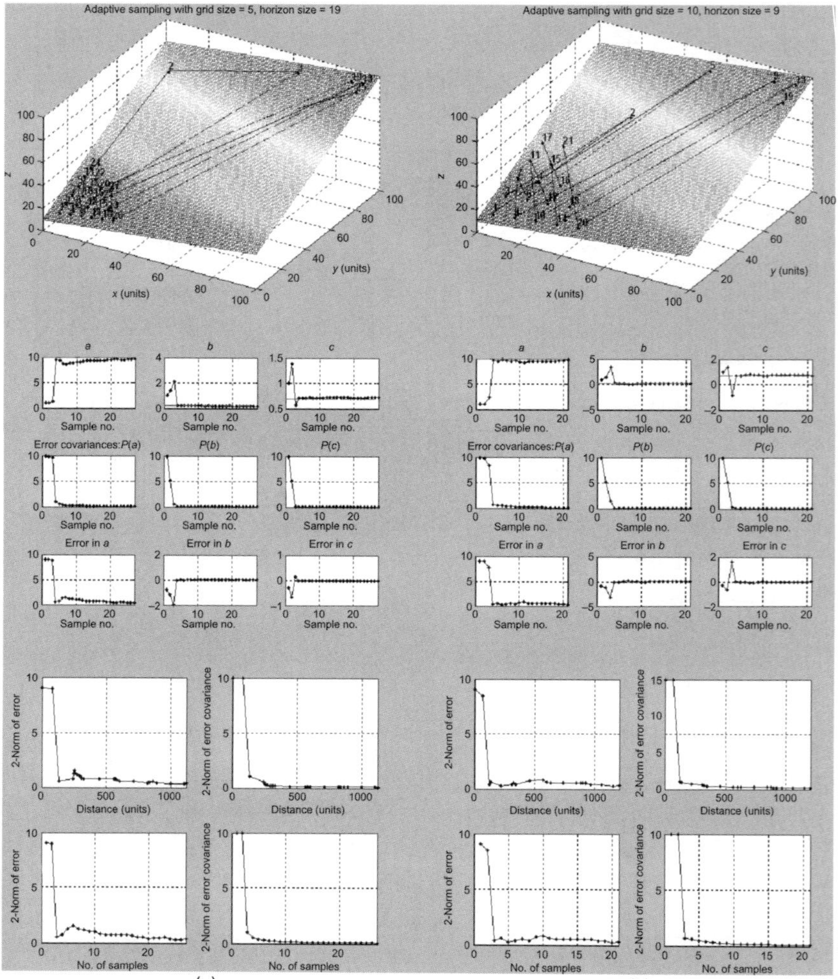


*Figure 3.7 Simulation results for adaptive sampling with (a) grid size 5 and horizon size 19 and (b) grid size 10 and horizon size 9*

- Measurement model:

$$
\begin{aligned}
\tilde{z}_k &= h_k(A_k) + v_k \\
\begin{bmatrix} \tilde{r_k} \\ \tilde{g_k} \\ \tilde{b}_k \end{bmatrix} &= \begin{bmatrix} r_{0k} + r_{1k}x + r_{2k}y \\ g_{0k} + g_{1k}x + g_{2k}y \\ b_{0k} + b_{1k}x + b_{2k}y \end{bmatrix} + v_k
\end{aligned} \tag{3.35}
$$

where $v_k \sim N(0, R)$, $x_0 \sim N(\bar{x}_0, P_0)$ and the model error covariance taken as $Q = 0$.

- Effect of system dynamics:

$$
\begin{aligned}
\hat{A}_{k+1}^{-} &= \hat{A}_k \\
P_{k+1}^{-} &= P_k
\end{aligned} \tag{3.36}
$$

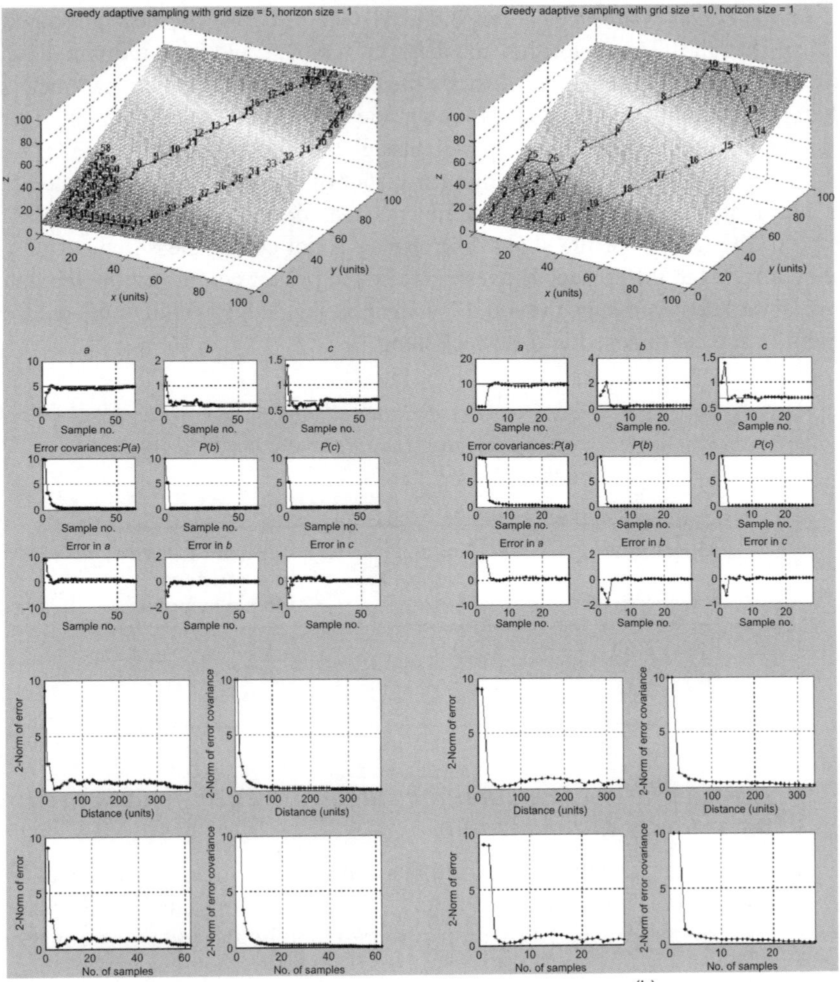


(a) (b)

*Figure 3.8 Simulation results for greedy adaptive sampling with (a) grid size 5 and horizon size 1 and (b) grid size 10 and horizon size 1*

- Effect of measurement:

$$
\begin{aligned}
H_{k+1} &= \begin{bmatrix} 1 & x & y & 0 & 0 & 0 & 0 & 0 & 0 \\ 0 & 0 & 0 & 1 & x & y & 0 & 0 & 0 \\ 0 & 0 & 0 & 0 & 0 & 0 & 1 & x & y \end{bmatrix} \\
K_{k+1} &= P_{k+1}^{-} H_{k+1}^{\mathrm{T}} (H_{k+1} P_{k+1}^{-} H_{k+1}^{\mathrm{T}} + R_{k+1})^{-1} \\
P_{k+1} &= (I_{12} - K_{k+1} H_{k+1}) P_{k+1}^{-} \\
\hat{A}_{k+1} &= \hat{A}_{k+1}^{-} + K_{k+1} \{ \tilde{z}_{k+1} - \hat{h}_{k+1} \} \\
&= \hat{A}_{k+1}^{-} + K_{k+1} \left\{ \begin{bmatrix} \tilde{r}_{k+1} \\ \tilde{g}_{k+1} \\ \tilde{b}_{k+1} \end{bmatrix} - \begin{bmatrix} \hat{r}_{0k+1}^{-} + \hat{r}_{1k+1}^{-} x + \hat{r}_{2k+1}^{-} y \\ \hat{g}_{0k+1}^{-} + \hat{g}_{1k+1}^{-} x + \hat{g}_{2k+1}^{-} y \\ \hat{b}_{0k+1}^{-} + \hat{b}_{1k+1}^{-} x + \hat{b}_{2k+1}^{-} y \end{bmatrix} \right\}
\end{aligned}
\tag{3.37}
$$

Experimental results in Figures 3.9–3.11 (see p. C2–4) are now presented to confirm that the GAS algorithm recovers the colour field in minimum time as shown in Table 3.3. The results illustrate the convergence of the field parameters to values close to nominal after successive samples are taken. The rate of convergence of the three algorithms can be compared through the speed of convergence for $r_0$, $g_0$ and $b_0$. The figures also show a camera snapshot of the actual field, estimated field and norm of the error in these coefficients that decreases with the number of samples. The reason for the slight discrepancy is not colour measurement or estimation errors, but differences between screen and printer colours on the laboratory floor. In these experiments, the robot localization is obtained through the overhead camera image, as opposed to dead reckoning.

*Table 3.3 Comparison of standard AS, greedy AS and raster scanning for linear colour field (experimental results)*

| **Algorithms** | **No. of samples ($n$)** | **2-Norm of 'error in coefficients' after $n$ samples** |
|---|---|---|
| Raster scan | 50 | 0.378 |
| Adaptive sampling | 15 | 0.365 |
| Greedy AS | 30 | 0.499 |

### *3.7.3 Kalman filter estimation for single Gaussian field with no uncertainty in localization*

We now consider sampling of a stationary non-linear field represented by a Gaussian with four unknown parameters, the centre, amplitude and variance $A = [a \quad \sigma \quad x_0 \quad y_0]$. This distribution is represented by the following equation:

$$Z = a \exp\left\{ -\frac{(x - x_0)^2 + (y - y_0)^2}{2\sigma^2} \right\} \tag{3.38}$$

#### 3.7.3.1 Minimization of uncertainty in single parameter

Four sampling possibilities exist in which the sampling criterion seeks to minimize uncertainties in estimates of individual parameters ($a$, $\sigma$, $x_0$, $y_0$). The numerical assumptions for simulation are as follows:

- Actual parameters: $[a \quad \sigma \quad x_0 \quad y_0] = [100 \quad 20 \quad 40 \quad 60]$.
- Initial parameters: $[a \quad \sigma \quad x_0 \quad y_0] = [105 \quad 15 \quad 45 \quad 65]$.
- Model error covariance: $Q = 0$.
- Measurement error covariance: $R = 1$.

We can observe that minimizing the error covariance of $a$, while disregarding the uncertainty in other parameters, leads to samples being taken at the current estimate of $(x_0, y_0)$ and in its neighbourhood. Similarly, minimizing the uncertainty in $\sigma$, $x_0$ and $y_0$ results in samples taken around $(x_0 \pm \sigma, y_0 \pm \sigma)$, $(x_0 \pm \sigma, y_0)$ and $(x_0, y_0 \pm \sigma)$, respectively. This idea is illustrated in Figure 3.12.

We also investigated the effect of changing grid sizes for the sampling simulations. We used the following numerical assumptions:

- Actual parameters: $[a \quad \sigma \quad x_0 \quad y_0] = [100 \quad 50 \quad 40 \quad 60]$.
- Initial parameter estimates: $[80 \quad 40 \quad 35 \quad 75]$.
- Initial error covariance: $P = \begin{bmatrix} 10 & 0 & 0 & 0 \\ 0 & 10 & 0 & 0 \\ 0 & 0 & 10 & 0 \\ 0 & 0 & 0 & 10 \end{bmatrix}$.
- Model error covariance: $Q = 0$.
- Measurement error covariance: $R = 1$.
- Convergence criteria: $||A - \hat{A}_{k+1}|| < 2.5$.

Simulation results are shown in Figures 3.13–3.15 for the three sampling schemes for individual parameter estimates, error covariance of individual parameters, error in individual parameter estimates, error in individual error covariance, 2-norm of parameter estimates and 2-norm of error covariance of parameters. The results are tabulated in Table 3.4 with cases on using different grid sizes also. As in the previous sections, GAS leads to a much smaller sampling distance, while AS requires the least number of samples.

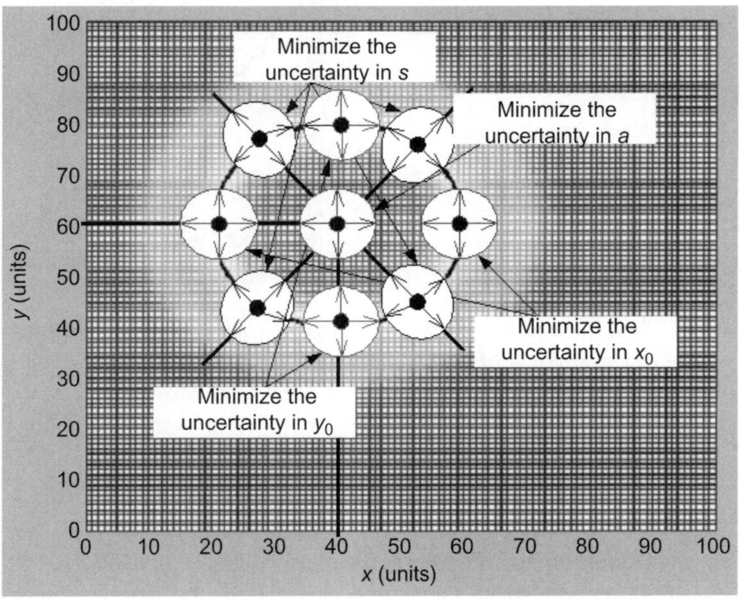


*Figure 3.12 Applying minimum error covariance criterion for different parameters results in specific areas of the Gaussian to be sampled*

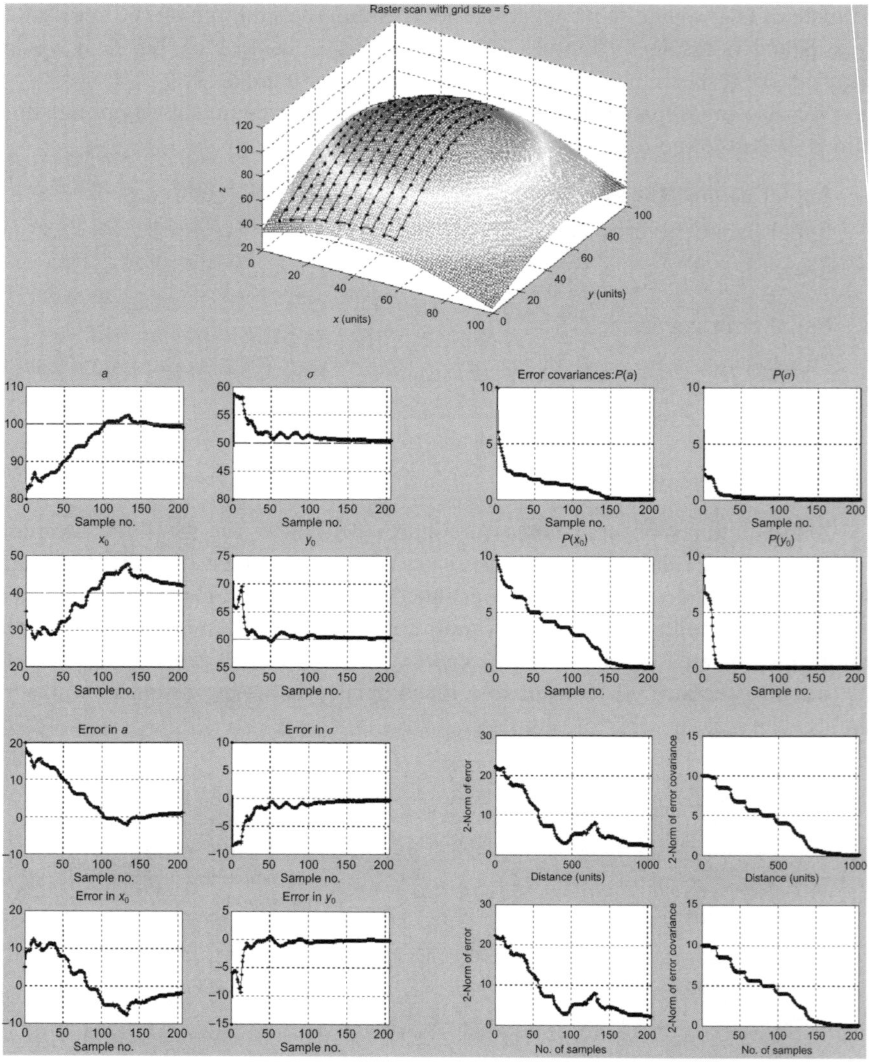


*Figure 3.13 Raster scanning of a Gaussian field with grid size 5 that takes 205 samples and a distance of 1020*

### *3.7.4 Simple Kalman filter estimation for linear field with uncertainty in localization*

In this case, in addition to the parameter state equation update $A_{k+1} = A_k$, the vehicle dynamics must also be taken into account. In the simple case when the vehicle motion can be represented by a single-particle equation with uncertainty $X_{k+1} = X_k + U_k + w_k$, and assuming that we use dead reckoning for localization and that the function $g(X_k)$ is linear, the kalman filter (KF) equations become

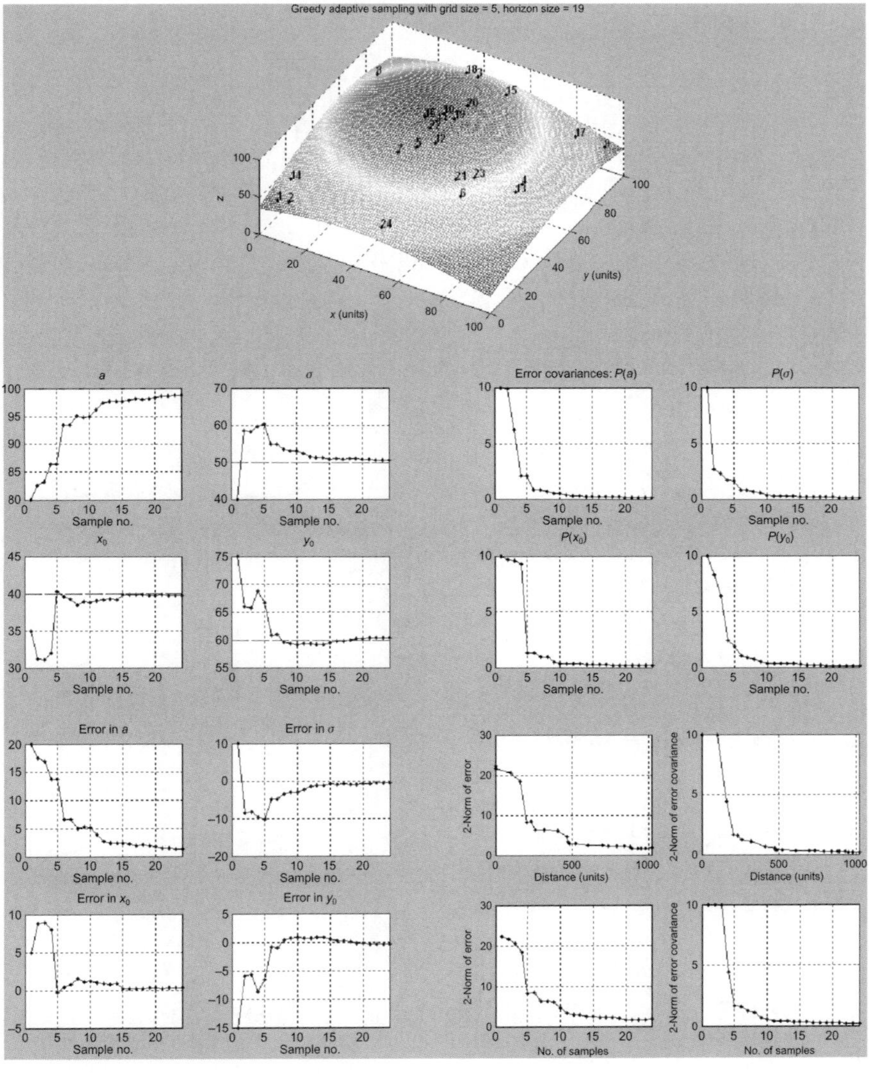


*Figure 3.14 Adaptive sampling of a Gaussian field with grid size 5 and horizon size 19*

$$
\begin{aligned}
\begin{pmatrix} X_{k+1} \\ A_{k+1} \end{pmatrix} &= \begin{pmatrix} X_k \\ A_k \end{pmatrix} + \begin{pmatrix} I_3 \\ 0 \end{pmatrix} U_k + \begin{pmatrix} w_k \\ 0 \end{pmatrix} = \begin{pmatrix} X_k \\ A_k \end{pmatrix} + BU_k + \vartheta_k \\
\begin{pmatrix} Y_k \\ Z_k \end{pmatrix} &= \begin{pmatrix} X_k \\ (1 \;\; g(X_k)^{\mathrm{T}})\, A_k \end{pmatrix} + \begin{pmatrix} \xi_k \\ \nu_k \end{pmatrix} = \begin{pmatrix} I_3 & 0 \\ 0 & (1 \;\; g(X_k)^{\mathrm{T}}) \end{pmatrix} \begin{pmatrix} X_k \\ A_k \end{pmatrix} + \lambda_k \\
X_{k+1}^{DR} &= X_k{}^{DR} + h(X_k{}^{DR}, u_k{}^{-})
\end{aligned}
\tag{3.39}
$$

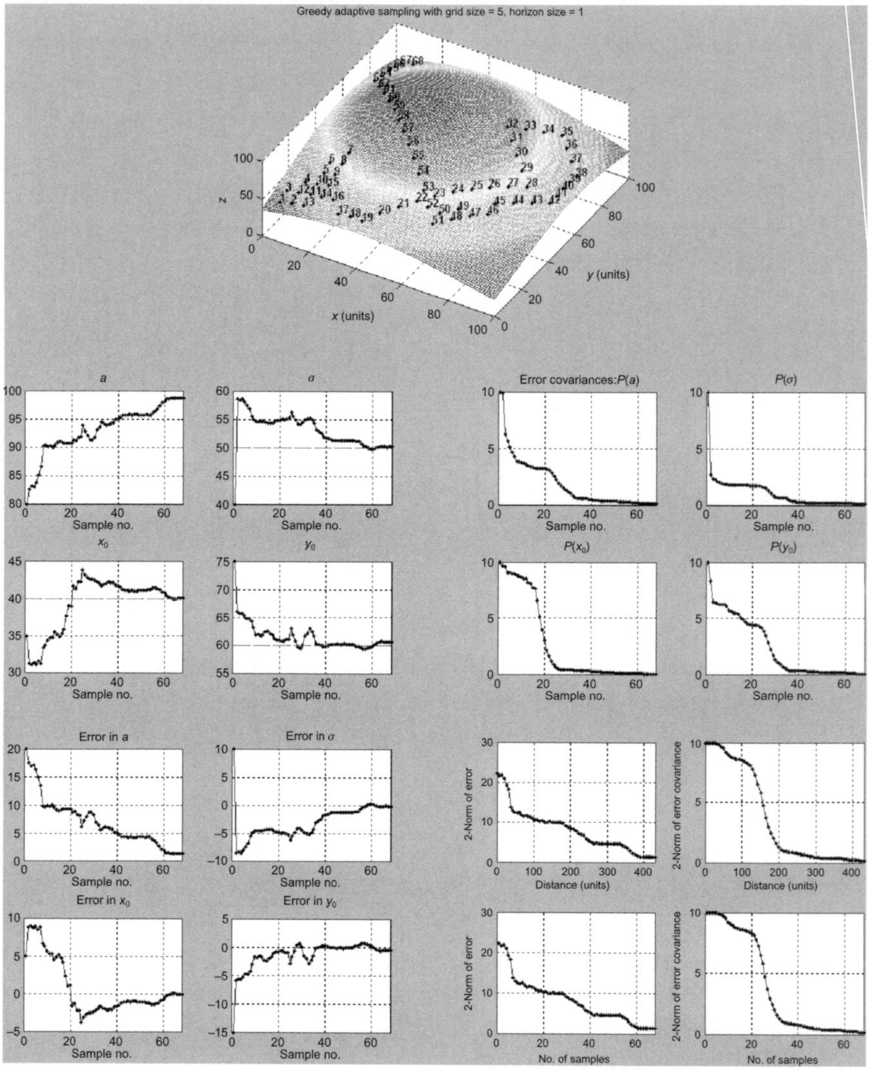


*Figure 3.15 Greedy adaptive sampling of a Gaussian field with grid size 5 and horizon size 1*

where $X_k^{DR}$ is the dead-reckoning estimate for the robot position computed after a command vector $u_k^-$ is applied to the robot to cause a predicted position change equivalent to $u_k$, and

*Table 3.4 Comparison of RS, GAS and AS for number of samples and distance when different grid sizes and horizon sizes are considered for sampling a Gaussian field*

| | **Grid size** | **Horizon size** | **No. of samples** | **Distance** |
|---|---|---|---|---|
| Raster scan | 5 | – | 205 | 1020 |
| | 10 | – | 96 | 950 |
| Adaptive sampling | 5 | 19 | 24 | 1022 |
| | 10 | 9 | 34 | 1835 |
| Greedy adaptive sampling | 5 | 1 | 68 | 430 |
| | 10 | 1 | 44 | 496 |

$$E[\vartheta_k \vartheta_k{}^{\mathrm{T}}] = Q = \begin{pmatrix} Q_1 & 0 \\ 0 & 0 \end{pmatrix}, \quad E[\lambda_k \lambda_k{}^{\mathrm{T}}] = R = \begin{pmatrix} R_1 & 0 \\ 0 & R_2 \end{pmatrix} \tag{3.40}$$

are the white noise covariances of the state and output.

The linear Kalman filter equations become

$$P_{k+1}^{-} = P_k + Q, \qquad \begin{pmatrix} X_{k+1}^{-} \\ \\ A_{k+1}^{-} \end{pmatrix} = \begin{pmatrix} \hat{X}_k \\ \\ \hat{A}_k \end{pmatrix} + BU_k$$

$$P_{k+1} = ((P_{k+1}^{-})^{-1} + G_{k+1}^{\mathrm{T}} R^{-1} G_{k+1})^{-1}$$

$$G_k = \begin{pmatrix} I_3 & 0 \\ \\ 0 & (1 \quad g(X_k^{\mathrm{T}})) \end{pmatrix}, \qquad \begin{pmatrix} X_{k+1}^{-} \\ \\ A_{k+1}^{-} \end{pmatrix} = \begin{pmatrix} X_k \\ \\ A_k \end{pmatrix} + BU_k$$

$$\begin{pmatrix} \hat{X}_{k+1} \\ \\ \hat{A}_{k+1} \end{pmatrix} = \begin{pmatrix} X_{k+1}^{-} \\ \\ A_{k+1}^{-} \end{pmatrix} + P_{k+1} G_{k+1}^{\mathrm{T}} R^{-1} \left[ \begin{pmatrix} Y_{k+1} \\ \\ Z_{k+1}^{-} \end{pmatrix} - G_{k+1} \begin{pmatrix} X_{k+1}^{-} \\ \\ A_{k+1}^{-} \end{pmatrix} \right] \tag{3.41}$$

If, on the other hand, the robot dynamics or the field is non-linear in parameters or in the robot state, (3.41) will become a non-linear Kalman filter (EKF).

In this section, we present experimental validation of the case when uncertainty in camera measurement is added to the model. Hence, there are now two measurements, namely $Y_k$ for localization and $Z_k$ for field measurement. The EKF equations consist of the following:

- System model:

$$\begin{pmatrix} X_{k+1} \\ A_{k+1} \end{pmatrix} = \begin{pmatrix} X_k \\ A_k \end{pmatrix} + \begin{pmatrix} B_k \\ 0 \end{pmatrix} U_k + \begin{pmatrix} w_k \\ 0 \end{pmatrix}$$

$$\begin{bmatrix} x \\ y \\ \theta \\ r_0 \\ r_1 \\ r_2 \\ g_0 \\ g_1 \\ g_2 \\ b_0 \\ b_1 \\ b_2 \end{bmatrix}_{k+1} = \begin{bmatrix} x \\ y \\ \theta \\ r_0 \\ r_1 \\ r_2 \\ g_0 \\ g_1 \\ g_2 \\ b_0 \\ b_1 \\ b_2 \end{bmatrix}_k + \begin{bmatrix} \frac{K_R r_R}{2}\cos(\theta_k) & \frac{K_L r_L}{2}\cos(\theta_k) \\ \frac{K_R r_R}{2}\sin(\theta_k) & \frac{K_L r_L}{2}\sin(\theta_k) \\ \frac{K_R r_R}{2L_b} & -\frac{K_L r_L}{2L_b} \\ 0 & 0 \\ 0 & 0 \\ 0 & 0 \\ 0 & 0 \\ 0 & 0 \\ 0 & 0 \\ 0 & 0 \\ 0 & 0 \\ 0 & 0 \end{bmatrix} \begin{bmatrix} \Delta\phi_R \\ \Delta\phi_L \end{bmatrix}_k + \begin{bmatrix} w_k \\ 0 \end{bmatrix} \tag{3.42}$$

where $w_k \sim N(0, Q)$ and $x_0 \sim N(\overline{x}_0, P_0)$.

- Measurement model:

$$\begin{bmatrix} \tilde{Y}_k \\ \tilde{Z}_k \end{bmatrix} = \begin{bmatrix} \tilde{x}_k \\ \tilde{y}_k \\ \tilde{\theta}_k \\ \tilde{r}_k \\ \tilde{g}_k \\ \tilde{b}_k \end{bmatrix} = \begin{bmatrix} x_k \\ y_k \\ z_k \\ r_{0k} + r_{1k}x_k + r_{2k}y_k \\ g_{0k} + g_{1k}x_k + g_{2k}y_k \\ b_{0k} + b_{1k}x_k + b_{2k}y_k \end{bmatrix} + \begin{pmatrix} \xi_k \\ \nu_k \end{pmatrix} \tag{3.43}$$

where $\xi_k \sim N(0, R_1)$ and $\nu_k \sim N(0, R_2)$.

- Effect of system dynamics:

$$\begin{aligned} A &= I_{12} \\ P_{k+1}^- &= P_k + Q_k \\ \hat{X}_{k+1}^- &= \hat{X}_k + \hat{B}_k U_k \\ \hat{A}_{k+1}^- &= \hat{A}_k \end{aligned}$$

$$\begin{bmatrix} \hat{x} \\ y \\ \hat{\theta} \\ \hat{r}_0 \\ \hat{r}_1 \\ \hat{r}_2 \\ \hat{g}_0 \\ \hat{g}_1 \\ \hat{g}_2 \\ \hat{b}_0 \\ \hat{b}_1 \\ \hat{b}_2 \end{bmatrix}_{k+1}^{-} = \begin{bmatrix} \hat{x} \\ y \\ \hat{\theta} \\ \hat{r}_0 \\ \hat{r}_1 \\ \hat{r}_2 \\ \hat{g}_0 \\ \hat{g}_1 \\ \hat{g}_2 \\ \hat{b}_0 \\ \hat{b}_1 \\ \hat{b}_2 \end{bmatrix}_{k} + \begin{bmatrix} \frac{K_R r_R}{2}\cos(\hat{\theta}_k) & \frac{K_L r_L}{2}\cos(\hat{\theta}_k) \\ \frac{K_R r_R}{2}\sin(\hat{\theta}_k) & \frac{K_L r_L}{2}\sin(\hat{\theta}_k) \\ \frac{K_R r_R}{2L_b} & -\frac{K_L r_L}{2L_b} \\ 0 & 0 \\ 0 & 0 \\ 0 & 0 \\ 0 & 0 \\ 0 & 0 \\ 0 & 0 \\ 0 & 0 \\ 0 & 0 \\ 0 & 0 \end{bmatrix} \begin{bmatrix} \Delta\phi_R \\ \Delta\phi_L \end{bmatrix}_k \tag{3.44}$$

- Effect of measurement:

$$\hat{H}_{k+1} = \frac{\partial \hat{h}_{k+1}}{\partial(\hat{X}_{k+1}^{-}, \hat{A}_{k+1}^{-})}$$
$$= \begin{bmatrix} 1 & 0 & 0 & 0 & 0 & 0 & 0 & 0 & 0 & 0 & 0 & 0 \\ 0 & 1 & 0 & 0 & 0 & 0 & 0 & 0 & 0 & 0 & 0 & 0 \\ 0 & 0 & 1 & 0 & 0 & 0 & 0 & 0 & 0 & 0 & 0 & 0 \\ \hat{r}_{1k+1}^{-} & \hat{r}_{2k+1}^{-} & 0 & 1 & \hat{x}_{k+1}^{-} & \hat{y}_{k+1}^{-} & 0 & 0 & 0 & 0 & 0 & 0 \\ \hat{g}_{1k+1}^{-} & \hat{g}_{2k+1}^{-} & 0 & 0 & 0 & 0 & 1 & \hat{x}_{k+1}^{-} & \hat{y}_{k+1}^{-} & 0 & 0 & 0 \\ \hat{b}_{1k+1}^{-} & \hat{b}_{2k+1}^{-} & 0 & 0 & 0 & 0 & 0 & 0 & 0 & 1 & \hat{x}_{k+1}^{-} & \hat{y}_{k+1}^{-} \end{bmatrix}$$

$$K_{k+1} = P_{k+1}^{-}\hat{H}_{k+1}^{\mathrm{T}}(\hat{H}_{k+1}P_{k+1}^{-}\hat{H}_{k+1}^{\mathrm{T}} + R_{k+1})^{-1}$$
$$P_{k+1} = (I_{12} - K_{k+1}\hat{H}_{k+1})P_{k+1}^{-}$$
$$\begin{pmatrix} \hat{X}_{k+1} \\ \hat{A}_{k+1} \end{pmatrix} = \begin{pmatrix} \hat{X}_{k+1}^{-} \\ \hat{A}_{k+1}^{-} \end{pmatrix} + K_{k+1}\left\{ \begin{bmatrix} \tilde{Y}_{k+1} \\ \tilde{Z}_{k+1} \end{bmatrix} - \hat{h}_{k+1} \right\} \tag{3.45}$$
$$= \begin{pmatrix} \hat{X}_{k+1} \\ \hat{A}_{k+1} \end{pmatrix} + K_{k+1}\left\{ \begin{bmatrix} \tilde{x}_{k+1} \\ \tilde{y}_{k+1} \\ \tilde{\theta}_{k+1} \\ \tilde{r}_{k+1} \\ \tilde{g}_{k+1} \\ \tilde{b}_{k+1} \end{bmatrix} - \begin{bmatrix} \hat{x}_{k+1}^{-} \\ \hat{y}_{k+1}^{-} \\ \hat{\theta}_{k+1}^{-} \\ \hat{r}_{0k+1}^{-} + \hat{r}_{1k+1}^{-}\hat{x}_{k+1}^{-} + \hat{r}_{2k+1}^{-}\hat{y}_{k+1}^{-} \\ \hat{g}_{0k+1}^{-} + \hat{g}_{1k+1}^{-}\hat{x}_{k+1}^{-} + \hat{g}_{2k+1}^{-}\hat{y}_{k+1}^{-} \\ \hat{b}_{0k+1}^{-} + \hat{b}_{1k+1}^{-}\hat{x}_{k+1}^{-} + \hat{b}_{2k+1}^{-}\hat{y}_{k+1}^{-} \end{bmatrix} \right\}$$

Through Monte Carlo experiments, the following uncertainties are determined and added to the EKF model:

(i) Camera accuracy error of zero mean and (2 in, 2 in, 2°) variance, meaning that our surrogate GPS measurements are accurate within 2 in and 2°.
(ii) Colour sensor uncertainty of zero mean and (0.01, 0.01, 0.01) variance, meaning that the colour sensor accuracy is 1 per cent RGB.

In Figure 3.16 (see p. C5), the lines in cyan colour show the dead-reckoning robot location based on the encoder counts. It is observed that the dead-reckoning error increases as the distance covered increases. At every sample, a colour measurement is taken by the robot, and the overhead camera finds the robot location, which is shown in red colour. Location information and colour sensor information are fused in the EKF model. The estimated robot locations are shown in black colour. After taking 25 samples, error covariance in $(x, y, \theta)$ drops from $(10^4, 10^4, 10^4)$ to (0.8239, 0.8257, 3.5078). The error covariance for the colour states drops to $(0.0034, 9.43 \times 10^{-7}, 1.58 \times 10^{-6}, 0.0034, 9.44 \times 10^{-7}, 1.58 \times 10^{-6}, 0.0034, 9.40 \times 10^{-7}, 1.58 \times 10^{-6})$.

The total norm of error in coefficients after 25 samples is 0.8741. Qualitatively, the AS algorithm recovers the printed image on the laboratory floor very well.

### *3.7.5 Kalman filter estimation for linear-in-parameters field with location measurement unavailable*

In previous cases, it is assumed that both the sensor node position measurement $(Y_k)$ and field variable measurement $(Z_k)$ can be used to estimate parameter states $(A_k)$ and position states $(X_k)$. In this case, we assume that the node position output is not present and we rely on field variable measurements alone, but we allow for the field to be linear in parameters but non-linear in the location state. Localization uncertainties are especially relevant in GPS-denied environments, but they can also be present in situations where local visual information is unavailable (for instance due to thick smoke), and GPS data rates are slow.

Here model parameter estimation for the field variable is integrated with estimation of the uncertainty in the mobile robot localization, and the overall covariance of the estimate is used for sampling. This way, localization uncertainty can be reduced by building accurate models of distributed fields and vice versa. For instance, if a robot is sampling an unknown field, but its location is accurately known, a distributed parameter field model can be constructed by taking repeated field samples. Later on, this field model can be used to reduce the localization error.

In the case of an $m$-dimensional linear-in-parameters field distribution:

$$g(X_k, A_k) = a_{k_0} + \sum_{i=1}^{m} g_i(X_k) a_{ki} = \begin{pmatrix} 1 & g_1(X_k)^{\mathrm{T}} \end{pmatrix} A_k \tag{3.46}$$

where $A_k$ is the parameter vector and is a non-linear basis $g_1$ for the field; the non-linear EKF equations can be written as

$$\begin{aligned} \begin{pmatrix} X_{k+1} \\ A_{k+1} \end{pmatrix} &= \begin{pmatrix} X_k \\ A_k \end{pmatrix} + \begin{pmatrix} I_3 \\ 0 \end{pmatrix} U_k + \begin{pmatrix} \omega_k \\ 0 \end{pmatrix} = \begin{pmatrix} X_k \\ A_k \end{pmatrix} + B_k U_k + \vartheta_k \\ Z_k &= (1 \;\; g_1(X_k)^{\mathrm{T}}) A_k + v_k \\ X_{k+1}^{DR} &= X_k{}^{DR} + h(X_k{}^{DR}, u_k{}^{-}) \end{aligned} \tag{3.47}$$

where

$$E\left[\vartheta_k \vartheta_k{}^{\mathrm{T}}\right] = Q = \begin{pmatrix} Q_1 & 0 \\ 0 & 0 \end{pmatrix}, \quad E[\lambda_k \lambda_k{}^{\mathrm{T}}] = R \tag{3.48}$$

The non-linear Kalman filter update equations become

$$\begin{aligned} P_{k+1}^{-} &= P_k + Q, \qquad \begin{pmatrix} \hat{X}_{k+1}^{-} \\ \hat{A}_{k+1}^{-} \end{pmatrix} = \begin{pmatrix} \hat{X}_k \\ \hat{A}_k \end{pmatrix} + B_k \left( X_{k+1}^{DR} - X_k^{DR} \right) \\ H_k &= \frac{\partial g_k}{\partial \left( \hat{X}_k^{-}, \hat{A}_k^{-} \right)} = \begin{pmatrix} \left( \hat{A}_k^{-} \right)^{\mathrm{T}} \begin{pmatrix} 0 \\ \dfrac{\partial g_1 \left( \hat{X}_k^{-} \right)}{\partial \hat{X}_k^{-}} \end{pmatrix} & \left( 1 \;\; g_1 \left( \hat{X}_k^{-} \right)^{\mathrm{T}} \right) \end{pmatrix} \\ P_{k+1} &= \left( \left( P_{k+1}^{-} \right)^{-1} + H_{k+1}^{\mathrm{T}} R^{-1} H_{k+1} \right)^{-1} \\ \begin{pmatrix} \hat{X}_{k+1} \\ \hat{A}_{k+1} \end{pmatrix} &= \begin{pmatrix} \hat{X}_{k+1}^{-} \\ \hat{A}_{k+1}^{-} \end{pmatrix} + P_{k+1} H_{k+1}^{\mathrm{T}} R^{-1} \left[ Z_{k+1} - g_{k+1} \left( \hat{X}_k^{-}, \hat{A}_k^{-} \right) \right] \end{aligned} \tag{3.49}$$

In this experimental validation for the case when only the colour measurement $Z_k$ is available, the colour sensor measurement is used for field estimation as well as for the robot location estimation. In other words, we are using the colour measurements to aid in the localization of the robot. The system model and update equations are the same as the previous case, but measurement equation are now as follows:

- Measurement model:

$$\begin{aligned} \tilde{z}_k &= h_k(X_k, A_k) + v_k \\ \begin{bmatrix} \tilde{r}_k \\ \tilde{g}_k \\ \tilde{b}_k \end{bmatrix} &= \begin{bmatrix} r_{0k} + r_{1k} x_k + r_{2k} y_k \\ g_{0k} + g_{1k} x_k + g_{2k} y_k \\ b_{0k} + b_{1k} x_k + b_{2k} y_k \end{bmatrix} + v_k \end{aligned} \tag{3.50}$$

- Effect of measurement:

$$
\begin{aligned}
\hat{H}_{k+1} &= \frac{\partial \hat{h}_{k+1}}{\partial(X^-_{k+1}, A^-_{k+1})} \\
&= \begin{bmatrix} \hat{r}^-_{1k+1} & \hat{r}^-_{2k+1} & 0 & 1 & \hat{x}^-_{k+1} & \hat{y}^-_{k+1} & 0 & 0 & 0 & 0 & 0 & 0 \\ \hat{g}^-_{1k+1} & \hat{g}^-_{2k+1} & 0 & 0 & 0 & 0 & 1 & \hat{x}^-_{k+1} & \hat{y}^-_{k+1} & 0 & 0 & 0 \\ \hat{b}^-_{1k+1} & \hat{b}^-_{2k+1} & 0 & 0 & 0 & 0 & 0 & 0 & 0 & 1 & \hat{x}^-_{k+1} & \hat{y}^-_{k+1} \end{bmatrix} \\
K_{k+1} &= P^-_{k+1}\hat{H}^{\mathrm{T}}_{k+1}(\hat{H}_{k+1}P^-_{k+1}\hat{H}^{\mathrm{T}}_{k+1} + R_{k+1})^{-1} \\
P_{k+1} &= (I_{12} - K_{k+1}\hat{H}_{k+1})P^-_{k+1} \\
\begin{pmatrix} \hat{X}_{k+1} \\ \hat{A}_{k+1} \end{pmatrix} &= \begin{pmatrix} \hat{X}^-_{k+1} \\ \hat{A}^-_{k+1} \end{pmatrix} + K_{k+1}\{\tilde{z}_{k+1} - \hat{h}_{k+1}\} \qquad (3.51) \\
&= \begin{pmatrix} \hat{X}_{k+1} \\ \hat{A}_{k+1} \end{pmatrix} + K_{k+1} \left\{ \begin{bmatrix} \tilde{r}_{k+1} \\ \tilde{g}_{k+1} \\ \tilde{b}_{k+1} \end{bmatrix} - \begin{bmatrix} \hat{r}^-_{0k+1} + \hat{r}^-_{1k+1}\hat{x}^-_{k+1} + \hat{r}^-_{2k+1}\hat{y}^-_{k+1} \\ \hat{g}^-_{0k+1} + \hat{g}^-_{1k+1}\hat{x}^-_{k+1} + \hat{g}^-_{2k+1}\hat{y}^-_{k+1} \\ \hat{b}^-_{0k+1} + \hat{b}^-_{1k+1}\hat{x}^-_{k+1} + \hat{b}^-_{2k+1}\hat{y}^-_{k+1} \end{bmatrix} \right\}
\end{aligned}
$$

Results indicate that the field parameters converge to values close to the actual values for ten samples, after which the norm of error starts to increase rather than decrease, due to the accumulation of dead-reckoning errors. Hence, it is a good idea to estimate the field for a few samples, and then stop the field estimation and switch instead to improve the localization using the estimated field. Here the location estimate stays close to the dead-reckoning estimate instead of getting closer to the true robot location (Figure 3.17 (see p. C6)). After taking 25 samples, the error covariance in $(x, y, \theta)$ drops from $(10^4, 10^4, 10^4)$ to (90.5653, 65.9856, 0.9902). The error covariance for the colour states drops to $(0.0058, 1.7359 \times 10^{-6}, 2.7066 \times 10^{-6}, 0.0079, 3.0633 \times 10^{-6}, 4.1337 \times 10^{-6}, 0.0066, 1.4988 \times 10^{-6}, 2.3830 \times 10^{-6})$.

The total norm of error in coefficients after 25 samples is 1.3568. In this case again, the AS algorithm recovers the printed image on the laboratory floor as shown in Figure 3.17.

#### 3.7.5.1 Simulations showing the effect of field measurement error on localization and field estimates

As expected, since only the field sensor measurement is used for localization, the increase in sensor noise adversely affects both the estimation of the field and the localization of the robot.

In Figure 3.18–3.21 (see p. C7–8), the lines in cyan colour show the dead-reckoning robot location calculated from a differential drive kinematics without using geometric and encoder uncertainties. The location estimate based on the colour sensor measurement EKF is shown in yellow, while the ‘true’ robot position calculated from differential drive kinematics with added geometric and encoder uncertainties is plotted in red. The ‘true’ sampling locations correspond to

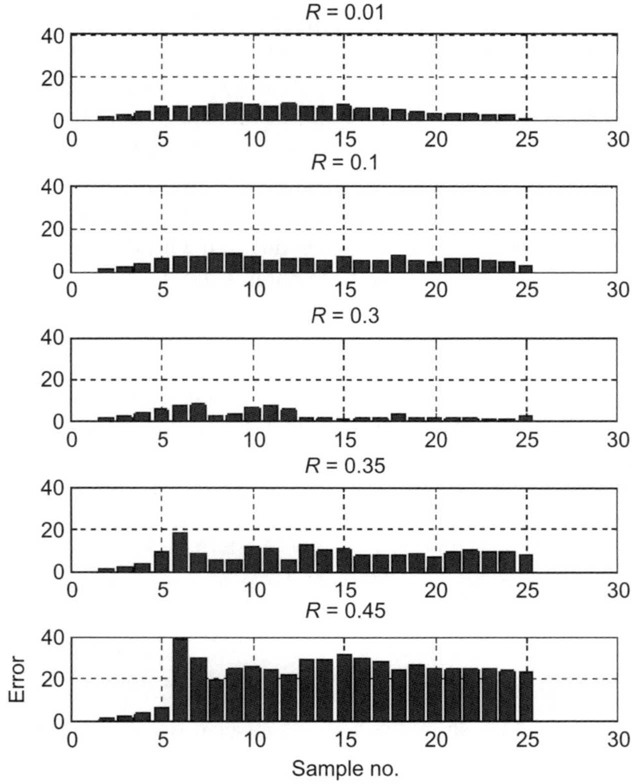


*Figure 3.22 Difference between the true and estimated robot locations versus the number of samples for different sensor measurement errors*

*Table 3.5 Effect of sensor measurement error on estimated field parameters*

| **Sensor measurement error (*R*)** | **Norm of error in estimated parameters (after 25 samples)** |
|---|---|
| 0.01 | 1.0240 |
| 0.1 | 1.0316 |
| 0.3 | 1.0911 |
| 0.35 | 1.1995 |
| 0.45 | 1.2462 |

measurements of position using the overhead camera. As expected, the dead-reckoning error increases as the robot covers more distance.

By comparing Figures 3.18–3.21, it is clear that the location estimate is close to the true location for small errors in colour sensor measurement. Figure 3.22 shows that large discrepancies in both the estimated field and robot position result for large measurement uncertainties. Table 3.5 shows the effect of sensor measurement error on estimated field parameters.

### 3.7.5.2 Simulation results showing the effect of localization uncertainty on field estimation for a Gaussian field

We now consider the case of sampling a stationary Gaussian field with four unknown parameters using a differential drive robot. Localization uncertainty is introduced by assuming different wheel radii and unequal wheelbases with a dead-reckoning localization algorithm (see Figure 3.23). Therefore, the robot ends up at locations different from where it thinks it is based on its dead-reckoning model.

The robot kinematic model is now included in the joint EKF, and both location states for the robot $X_k = [x_k \; y_k]$ and field parameters $A = [a \; \sigma \; x_0 \; y_0]$ need to be estimated.

Following numerical assumptions were used in the simulation:

- Actual field parameters: $[a \quad \sigma \quad x_0 \quad y_0] = [100 \quad 50 \quad 40 \quad 60]$.
- Initial parameter estimates: $[80 \quad 30 \quad 45 \quad 55]$.
- Final parameter estimate: $[99 \quad 50 \quad 42 \quad 57]$.
- States to estimate: $[a \quad \sigma \quad x_0 \quad y_0 \quad x \quad y]$.
- Initial error covariances: $P_{0a} = 100$, $P_{0\sigma} = 100$, $P_{0x0} = 100$, $P_{0y0} = 100$, $P_{0x} = 100$ and $P_{0y} = 100$.
- Model error covariances: $Q_a = 0$, $Q_\sigma = 0$, $Q_{x0} = 0$, $Q_{y0} = 0$, $Q_x = 1$ and $Q_y = 1$.
- Measurement error covariance: $R = 1$.
- The robot wheel radii are different: $r_1 = 1$ and $r_r = 1.05$.

Figures 3.24 and 3.25 show the actual Gaussian field, the initial estimate, the path taken by the robot during sampling and the reconstructed field after 100 samples of GAS. Figure 3.26 plots the 2-norm of the error between the actual robot position (based on the actual wheel radii), the dead-reckoning estimate (based on incorrect values in the model) and the robot location estimate as computed by the joint EKF. It can be observed from Figure 3.26 that initially the dead-reckoning error (in dotted lines) is small, but it grows as time passes. Also, initially the estimated location based on field measurements is not very good

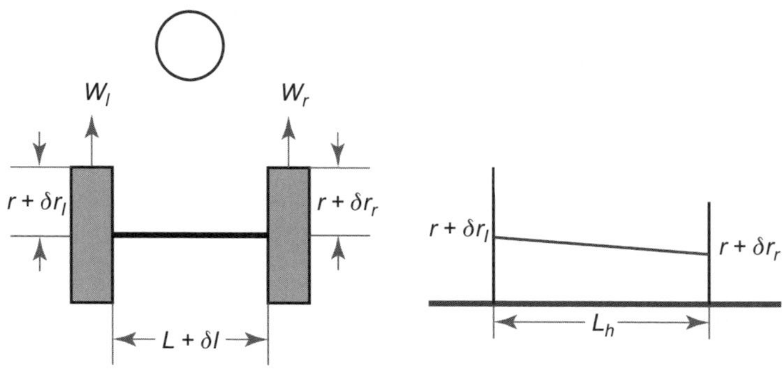


*Figure 3.23 Differential drive mobile robot with uncertainty in wheel radii and axle lengths*

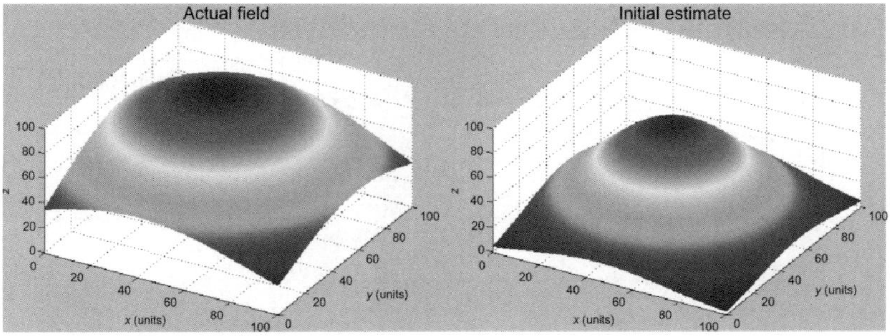


*Figure 3.24 The actual field (left) and the initial estimate (right)*

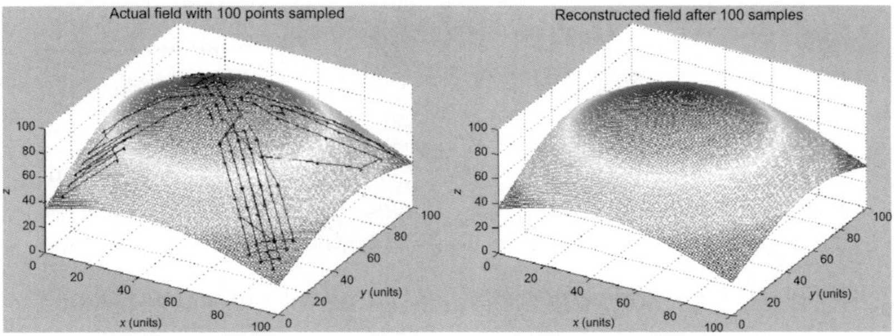


*Figure 3.25 The actual field with points sampled using GAS (left) and the reconstructed field after 100 samples (right)*

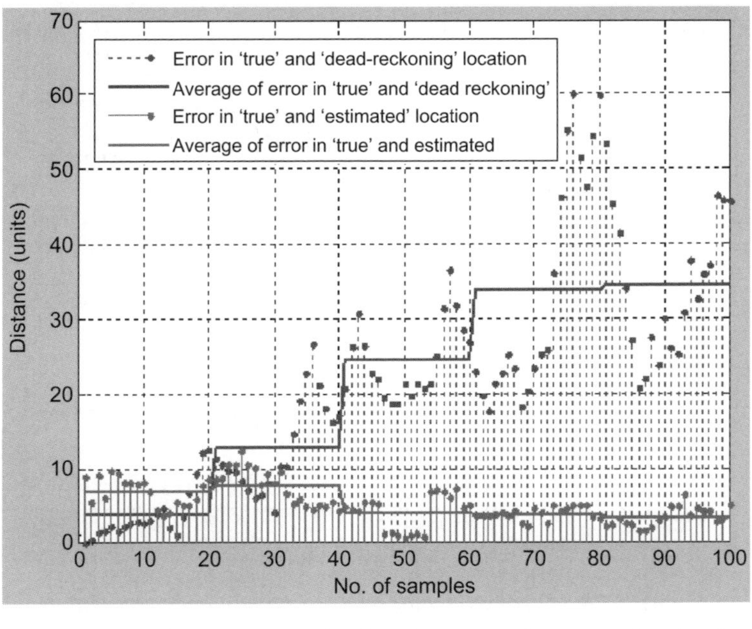


*Figure 3.26 Error between true, dead-reckoning and estimated robot locations*

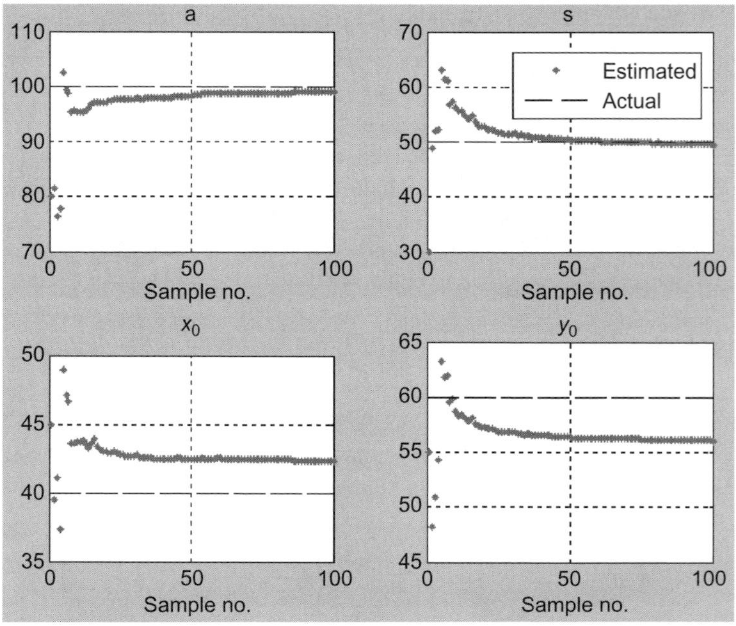


*Figure 3.27 Convergence of field parameters to actual values*

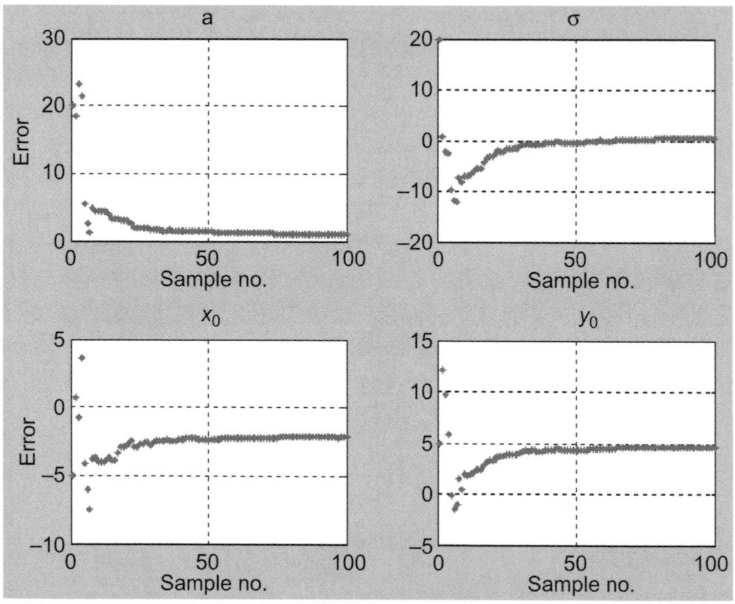


*Figure 3.28 Error in actual and estimated parameter values*

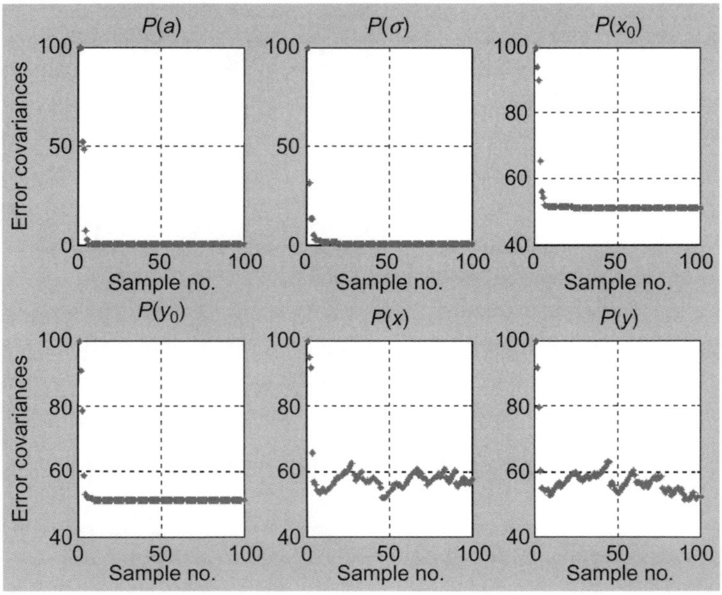


*Figure 3.29 Error covariance of location states and field parameters*

(in solid lines). However, while the dead-reckoning error keeps accumulating, the robot localization error stays bounded with increasing number of samples because as more samples are taken the algorithm has better knowledge of the field, and this helps localize the robot. A windowed average for the parameter estimates is taken every 20 samples, and the trends are shown by solid horizontal lines in Figures 3.27–3.29.

## 3.8 Summary

In summary, this chapter introduced the use of the EKF for sampling of parametric fields, discussed how to incorporate measurement and robot location uncertainty to drive the robot to sample at a location that minimizes both uncertainties, discussed the use of RBF neural networks to approximate complex fields, and the heuristics that can be used to reduce the computational complexity of the search necessary to select sampling locations for the robots.

# Chapter 4
# Case study: application to forest fire mapping

This chapter applies the adaptive sampling (AS) methods developed in the previous chapters to the application of forest fire mapping. The spread of forest fires is modelled as a parametric field, and the AS algorithms are made use of to estimate the fire field. Extensive simulations are carried out to validate the proposed algorithms.

The chapter is organized as follows: section 4.1 presents the two parametric models used to describe the spread of forest fires, section 4.2 discusses the parameterization of the field by interpreting remote sensing images, section 4.3 presents the formulation for extended Kalman filter (EKF)-based AS algorithm for spatio-temporal distributions and the multi-scale algorithm for mapping of forest fires using AS, and section 4.4 discusses potential field–based path planning for robots navigating through the estimated fire field. Finally, section 4.5 presents the simulation results for complex fields including the forest fires.

## 4.1 Parametric description of forest fire spread

The physics of the fire behaviour, its spread and immediate effects have been studied extensively [89–92]. Fuel, weather and topography are the key considerations in the spreading of fire. Most common are the semi-empirical fire modelling that uses Rothermel's equation for calculation of local rate of fire spread [90,92], Huygens' principle for modelling the shape of fire front [90,92] and the use of discrete event (DE) cellular automata (CA) models [93–96]. Furthermore, neural network and other classification schemes have been used for some time for supervised classification of remote sensing data, especially in applications of urban planning and atmospheric modelling, for instance, in order to classify the area as water, forest, wetlands, lakes, etc. [94]. Since the fire spread phenomenon is too complex to understand, and the effect of each and every variable is difficult to predict, a neural network can be used as a 'black box' for measuring the influence of the independent variables. For instance, representing the fire spread as a parametric time-varying sum of several Gaussians makes for an attractive approximation.

In this book, complex forest fire distributions are modelled through the use of radial basis function (RBF) neural networks. RBF parametric models resulting from a low-resolution satellite image, for instance, serve as good low-resolution initial approximations for the fire field. When combined with a high-resolution AS strategy, the errors introduced by low-resolution sampling and training are reduced. Sampling is done heuristically by mobile robots (agents) that search in the vicinity of the current location for future sampling locations, minimizing field parameter uncertainty. This results in a quick, high-resolution reconstruction of the field at the end of the sampling process. Furthermore, the parametric description of the field can be used along with the robot dynamic model to reduce the localization uncertainty of the agents. This may not be needed in scenarios where GPS measurements are available, but it is critical in GPS-denied environments, such as underwater or inside buildings.

The validity and performance of the sampling algorithm is tested here with increasingly complex fire spread scenarios. First, a simple time-varying elliptical forest fire spread model is considered that has been fitted to empirical data. This model considers a decreasing-intensity Gaussian distribution from the head of fire to the fire tail. The model is accurate at the beginning of the fire spread, and less accurate as time goes on. In later sections of this chapter, we consider more realistic fire spread models simulated by CA, using discrete rules to decide the burning of certain cells based on conditions at neighbouring cells. A low-resolution satellite image is used to approximate this field by an RBF neural network training algorithm that also acts as a classifier. The objective of the classification is to introduce more neurons in high-variance areas of the field, and fewer neurons in low-variance areas. Training of the neural net is only needed infrequently, and it need not be very accurate.

In addition to sampling and reconstruction of the fire field, it is sometimes necessary to direct the robots or human personnel through this field and avoid dangerous locations, e.g. 'hot zones'. This type of navigation can be accomplished using potential fields [97], and the estimated fire field model itself can serve as the navigation potential. Extensive research has been done in the area of path planning using potential fields for robotic vehicles [98]. Obstacle avoidance or goal attainment schemes often use penalty functions to bend feasible paths around obstacles as introduced by Krogh [99] and Kathib [100]. As the fire intensity field is generated via sampling, we can use it to generate fire-safe trajectories to and from rescue locations for firefighter human crews.

The sampling and navigation algorithms presented in this chapter will also be validated using extensive realistic, time-varying forest fire simulation models.

The fire spread phenomenon is highly complex, and existing mathematical models of its spread require extensive computing resources. Regenerating time-varying fire spread models might be adequate to run in batch mode, but they are difficult to use in real-time firefighting scenarios. However, these models help in understanding the influence of different factors on fire propagation without the need for an analytical solution.

Existing fire spread models are divided into three classes: empirical, semi-empirical and physical models. Using semi-empirical models is particularly interesting for our sampling approach because they use empirical measurements in addition to a physical description of the forest, hence providing a parametric space–time distributed field whose parameters can be estimated by AS. By contrast, physical models are too simplistic, while empirical models involve too many parameters for the EKF. The most common fire modelling approaches include semi-empirical models using Rothermel's equation for the calculation of local rate of fire spread and Huygens' principle for modelling the shape of fire front and DE CA models. Several powerful software applications such as FARSITE, FireGIS, Fire!, SPREAD, etc. are available to demonstrate the fire spread.

Many factors influence fire behaviour, with fuels, weather and topography being the primary factors. Parametric models have been empirically fitted to observe data using four primary inputs: fuel type, fuel moisture, wind and slope. Second-order variables such as temperature, humidity, shading and shelter operate through one of the four primary inputs. The effect of the primary inputs is described as follows:

- *Slope*: Fire can spread significantly faster up a slope than on level terrain in the same fuels. In many cases, topographic maps are very helpful for understanding the fire spread.
- *Wind*: Wind speed and direction are the most critical factors required for predicting fire behaviour.
- *Fuel*: Fuel types are lumped together into different models. Fuel loading, fuel depth, fuel particle density, head content of fuel, moisture of extinction and surface area to volume ratio are the factors used to further describe it.
- *Fuel moisture*: Fuel moisture is an expression of the amount of water in the fuel component. It determines both fire intensity and the heat required to bring the fuel ahead of spreading fire up to the ignition temperature.

### 4.1.1 Simple elliptical fire spread model

Given homogenous fuel and weather conditions, and assuming a constant, moderate wind, a fire growing from an ignition point is initially close to an elliptical shape. Strong winds or steep slopes can elongate the shape, but it remains consistent until fuels or wind change. The spread distance is the product of projection time and the rate of spread.

The most common simulation model for forest fire growth from point sources uses the Huygens' principle, which considers the elliptical nature of fire growth with $(a, b)$ as the minor and major axes, respectively, and $(x_s, y_s)$ as the distance from the centre of the fire ellipse to the fire source. Figure 4.1 depicts the elliptical spread of a fire starting at an ignition point. Since this function is a constrained Gaussian, it will later be used as an RBF function with constraints. For a typical case, the dimensions of the elliptical axis can be calculated by the following equations [90,92]:

$$
\begin{aligned}
& a = 0.5\frac{R + R/HB}{LB}, \qquad b = \frac{R + R/HB}{2}, \qquad c = b - \frac{R}{HB} \\
& LB = 0.936\,\mathrm{e}^{0.2566U} + 0.461\,\mathrm{e}^{-0.1548U} - 0.937 \qquad (4.1) \\
& HB = \frac{LB + (LB^2 - 1)^{0.5}}{LB - (LB^2 - 1)^{0.5}}
\end{aligned}
$$

where $R$ is the fire spread rate in ft/min, $LB$ is the length to breadth ratio, $HB$ is the head to back ratio and $U$ is the wind speed.

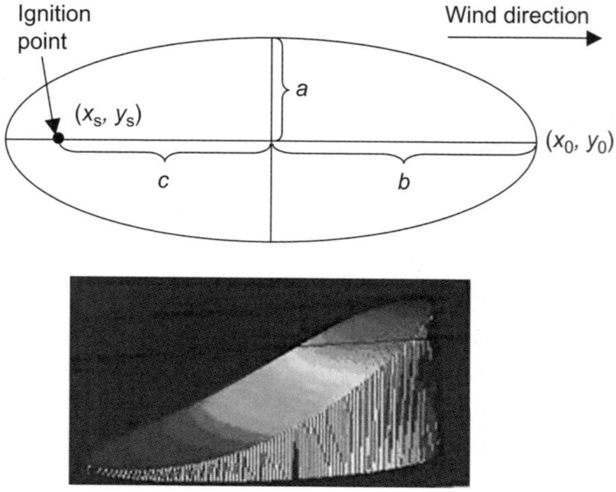


*Figure 4.1 Elliptical fire growth model represented by an elliptically constrained RBF function*

The variable that will later need to be mapped is the temperature of the fire field, a Gaussian distribution only defined in an elliptical region (Figure 4.2). It is further assumed that the maximum fire temperature $I$, variance $\sigma$ and $y_0$ are time invariant, while the firehead ($x_0(t)$, $y_0(t)$) location is proportional to the fire spread rate that in turn is proportional to the wind speed. If $y_0(t) = C$ (e.g. fire spreads horizontally), then the fire intensity is given by

$$g(X,A,t) = \begin{cases} I \exp\left[-\dfrac{(x - x_0(t))^2 + (y - y_0)^2}{2\sigma^2}\right], & \dfrac{(x - x_0(t) + b(t))^2}{b(t)^2} + \dfrac{(y - y_0)^2}{a(t)^2} \le 1 \\ 0, & \text{otherwise} \end{cases} \tag{4.2}$$

where

$$x_0(t) = x_s + b(t) + c(t) \tag{4.3}$$

If wind is blowing at an angle $\phi$ to the $x$-axis, the new parametric equation of ellipse can be derived by a rotation around the fire source ($x_s$, $y_s$) by angle $\phi$. In this case, (4.2) becomes

$$g(X_{new}, A, t) = I \exp\left[-\frac{||X_{new} - X_s||^2 + ||X_s - X_0||^2 + 2(X_{new} - X_s)^T R_\varphi (X_s - X_0)}{2\sigma^2}\right] \tag{4.4}$$

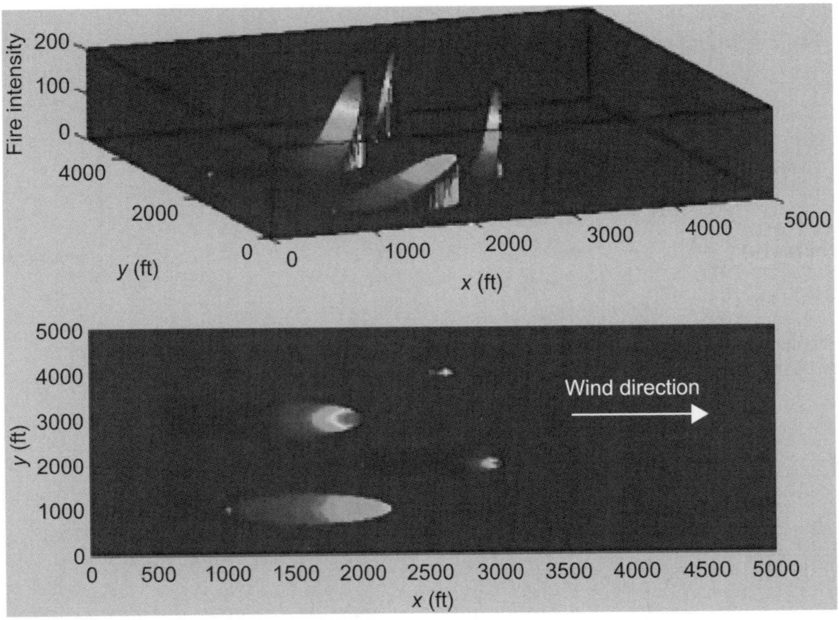


*Figure 4.2 Four fire ellipses originated at different points and with different spread rate, peak intensity and variance. The head of fire is the mean of the Gaussian distribution*

where $R_{\varphi}{}^{T}$ is a 2 × 2 rotation by angle $\phi$, and

$$X_{new} = [x_{new} \ y_{new}], \qquad X_0 = [x_0 \ y_0], \qquad X_s = [x_s \ y_s] \tag{4.5}$$

### *4.1.2 Complex cellular automata–based discrete event model*

The Huygens' fire spread model is quite simple, but it is quite reasonable for short intervals of time after ignition. Because it assumes uniform, continuous fuel, uniform wind velocity throughout the burning area and flat terrain, the model becomes inaccurate if the fire spreads over large distances. CA are very good for modelling and simulating complex dynamic systems whose evolution depends on the state of the current cell, neighbouring cells, wind, slope and fuel. This scheme can also recover the Huygens' fire spread model discussed in the previous section at a cellular level. The fire spread region is divided into cells, as shown in Figure 4.3, and the fire propagates from cell to cell governed by DE rules that iterate the temperature at the next sample time as a function of the temperature of its cell neighbours and of several local fire parameters. The transition rules can be written as

$$T_{i,j}^{k+1} = f(T_{i,j}^{k}, T_{i,j,1}^{k}, \ldots, T_{i,j,4}^{k}, \text{wind, slope, fuel}) \tag{4.6}$$

where $T_{i,j}^{k}$ are temperatures of the current cell and its neighbours.

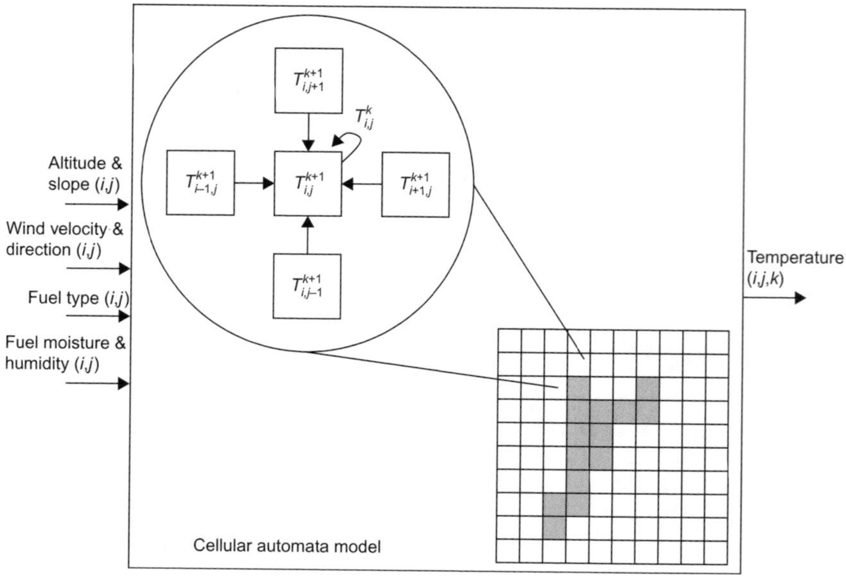


*Figure 4.3 CA model inputs, outputs and transition rules*

Each cell is considered to be in one of the following four states: *inactive, unburned, burning* and *burned*, as shown in Figure 4.4. The fire starts from a hot spot in an elliptical shape in the sense that each burning cell generates an ellipse with focus on the cell centre. Depending on fuel type, moisture, slope, wind and the state of the surrounding cells, each may have different rate of increase of temperature ($\alpha$), maximum temperature ($T_{max}$), ignition temperature ($T_{ignition}$), rate of temperature decay ($\beta$), time threshold for switching to unburned state ($t_{thresh}$) and ignition time ($t_{ignition}$).

The following set of equations governs the temperature variation with respect to time, as suggested by Reference 96:

$$T(t) = \begin{cases} T_a, & t \leq t_{th} \\ T_a + e^{(t-t_{th})\alpha}, & t_{th} < t \leq t_{max} \\ T_{max}, & t_{max} < t \leq t_{max} \\ T_a + (T_{max} - T_a)e^{(t_{maxf}-t)\beta}, & t > t_{maxf} \end{cases} \tag{4.7}$$

The rules can be simply defined as follows:

- A cell remains in *inactive* state until it is outside the *burning* ellipse. The state switches to *unburned* state when one of its neighbours starts *burning*.
- Once the cell switches from *inactive* state to *unburned* state, its temperature starts to rise at a rate of $\alpha$, which depends on the fuel availability, wind and terrain.
- The cell starts *burning* after it reaches the ignition temperature. The combustion duration and rate of temperature decay $\beta$ depend on fuel, wind and terrain.

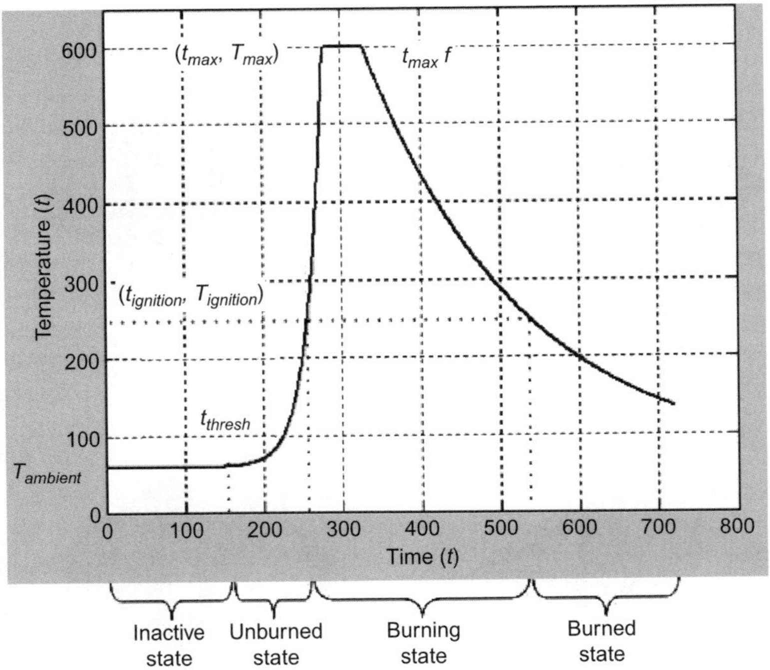


*Figure 4.4 Cell temperature variations with respect to time*

Figure 4.5 shows a MATLAB simulation of a fire generated from the hot spot at (244, 127) using CA at time instants $t = 2$, 3, 4 and 5 min. The number of pixels for the fire field was $x \times y = 300 \times 300$. The fire is circular in shape because slope variation and wind direction are not considered. But because of the different fuel present in each cell, the temperature is different and keeps increasing. Figure 4.6 (see p. C9) shows the fire at time $t = 20$, 40, 60 and 80 min in (a), (b), (c) and (d) after some time has elapsed and most of the area with significant fuel content is burning. In this chapter, these fire field models are sampled and reconstructed using the algorithms discussed in Chapter 3.

## 4.2 Neural network for parameterization

Digital remote sensing images of forests can be acquired from field-based, airborne and satellite platforms. Imagery from each platform can provide a data set with which forest analysis and modelling can be performed. Airborne images typically offer greatly enhanced spatial and spectral resolution over satellite images (Figure 4.7). In addition, airborne images provide greater control over capturing images only from specific areas of interest and at different altitudes. In section 3.3, we discussed the parameterization of complex time-varying fields that is applied here to a fire spread model using RBF neural nets. We assume that low-resolution images are available from overhead imagery at infrequent time intervals.

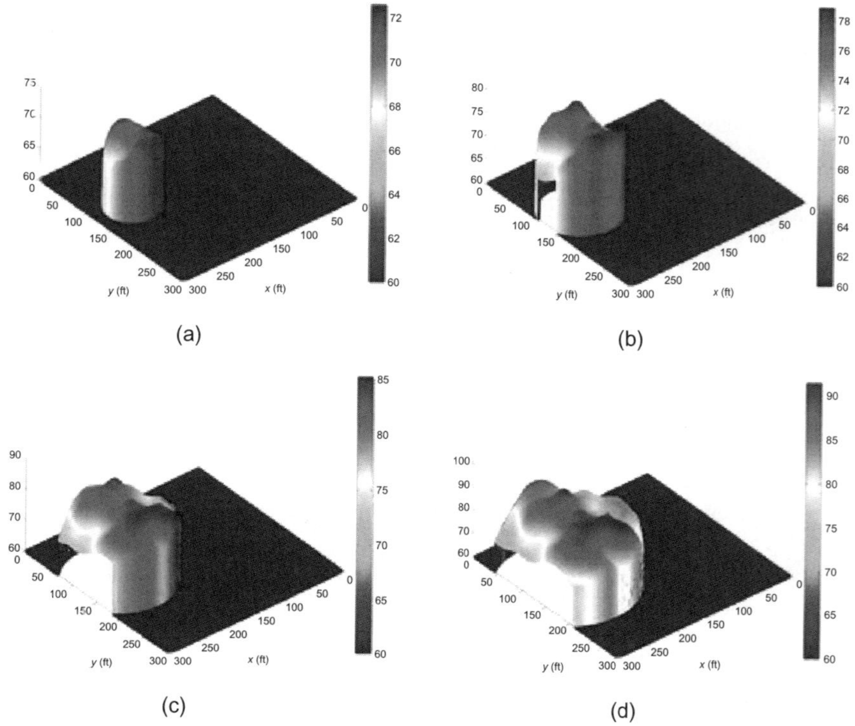


*Figure 4.5 Fire at time t = 2, 3, 4 and 5 min shown in (a), (b), (c) and (d), respectively*

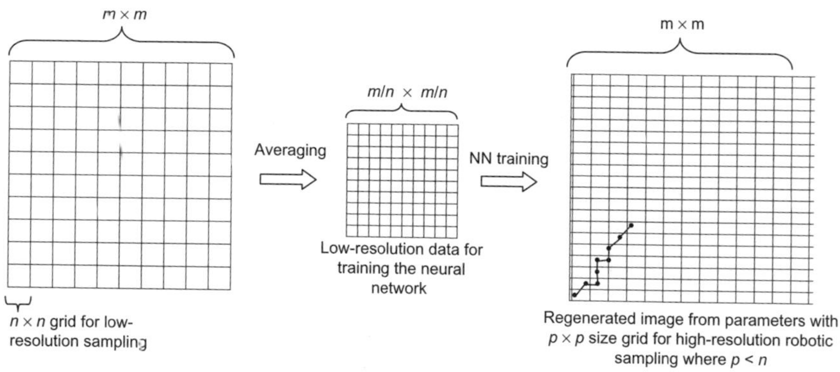


*Figure 4.7 Change in spatial resolutions for multi-scale sampling*

In the case of fire distribution, an exact non-linear model description is unattainable due to the high level of complexity. Instead, a parameterized approximation of the field is used, which is acquired by means of a neural network. In order to obtain an initial approximation of the field, a neural network is trained with a low-resolution 'fire field image'. Training is done at lower rates in this EKF estimation process (e.g. of sampling with robots). The network is presented with training pairs, which in our case are temperatures at different locations taken from a low-resolution infrared image. Therefore, we roughly approximate the complex spatio-temporal field with a sum-of-Gaussian parametric field by means of the universal approximation theorem.

## 4.3 EKF adaptive sampling of spatio-temporal distributions using mobile agents

The presented non-linear EKF approach is an efficient framework for combining the uncertainty in robot localization with errors in field sensor measurements. State-measurement minimization of the EKF covariance matrix norm is used to achieve effective AS using a variety of mobile robotic platforms including underwater and indoor vehicles. Localization uncertainties are especially relevant in GPS-denied environments, but they can also be relevant to fire fields in situations where local visual information is unavailable (for instance due to thick smoke) and GPS data rates are slow. Time-varying complex models are considered here stemming from realistic fire spreading simulations of the previous sections.

Model parameter estimation for the field variable (fire temperature) is integrated with estimation of the uncertainty in the mobile robot localization, and the overall estimate covariance is used for sampling. This way, localization uncertainty can be reduced by building accurate models of distributed fields and vice versa. For instance, if a robot is sampling an unknown field, but its location is accurately known, a distributed parameter field model can be constructed by taking repeated field samples. Later on, this field model can be used to reduce the localization error of the robot.

The multi-agent AS problem considered can be described as follows:

*Assumptions*:

(i) *A non-linear spatio-temporal field variable is described via a parametric approximation $Z = Z(A, X, t)$ depending on an unknown parameter vector $A$, position vector $X$ and time $t$.*

(ii) *$N$ robotic vehicles (agents) sample the field with localization and sensing uncertainty in order to obtain higher resolution estimates of the field, and also to improve their own location estimates.*

(iii) *The number of field parameters ($M$) and their initial guess is based on a hypothesis originating from prior knowledge of the field consistent with a low-resolution image of the entire field.*

In this section, the mapping of a slow time-varying complex forest fire field is considered, with measurements performed at different rates. For sampling slow-varying fields, the time update is performed at a slower rate compared with the measurement update. The time update (3.15) is performed at a sampling rate of $T_d$

by high spatial infrared imaging with an uncertainty represented by the process covariance noise $Q_{Td}$. The field evolution is measured by the difference $(\tilde{U}_{Td+1} - \tilde{U}_{Td})$ between these consecutive measurements. High-rate field measurement updates are then done by robotic sampling, with uncertainty represented by covariance noise $R_{2k}$.

Assuming that the $i$th robot location $Y_k^i$ is measured using some absolute localization scheme such as GPS, and if, for simplicity, we ignore the robot dynamics and localization uncertainty, we can write the state and output equations as

$$\begin{aligned} A_{k+1} &= A_k + (\tilde{U}_{Td+1} - \tilde{U}_{Td}) + \alpha_{Td+1} \\ \tilde{U}_{Td+1} &= U_{Td+1} \\ Y_k^i &= X_k^i \\ Z_k &= g(Y_k^i, A_k) + \nu_k \end{aligned} \tag{4.8}$$

Therefore, the EKF equations become

$$\begin{aligned} P_{k+1}^- &= P_k + Q_{Td+1}, \qquad \hat{A}_{k+1}^- = \hat{A}_k + (\tilde{U}_{Td+1} - \tilde{U}_{Td}) \\ H_k &= \frac{\partial g_k}{\partial (Y_k^i, \hat{A}_k^-)} \\ P_{k+1} &= ((P_{k+1}^-)^{-1} + H_{k+1}^{\mathrm{T}} R^{-1} H_{k+1})^{-1} \\ \hat{A}_{k+1} &= \hat{A}_{k+1}^- + P_{k+1} H_{k+1}^{\mathrm{T}} R^{-1} [Z_{k+1} - g_{k+1}(Y_{k+1}^i, \hat{A}_{k+1}^-)] \end{aligned} \tag{4.9}$$

where $Q_k = Q_{Td+1}$ and $R_k = R_{2k}$.

### 4.3.1 Formulation for elliptically constrained single Gaussian time-varying field

In this section, a time-varying field is assumed where the input $U_{2k}$ depends on the measured velocity of spread (wind velocity). The Kalman filter equations are used to estimate the peak intensity, variance and mean of the time-varying forest fire field. Using equation, the fire model equation can be written as

$$\begin{aligned} & A_{k+1} = A_k + B_{2k} U_{2k} + \alpha_k \\ & \begin{bmatrix} I \\ \sigma \\ x_0 \\ y_0 \end{bmatrix}_{k+1} = \begin{bmatrix} I \\ \sigma \\ x_0 \\ y_0 \end{bmatrix}_k + \begin{bmatrix} 0 \\ 0 \\ r_{k+1} \Delta t_{k+1} \\ 0 \end{bmatrix} + \begin{bmatrix} 0 \\ 0 \\ \alpha \\ 0 \end{bmatrix} \\ & \alpha \sim N(0, Q_2) \end{aligned} \tag{4.10}$$

where $\Delta t_{k+1}$ is the time from sample $k$ to $k+1$, $x_0$ is a continuous function that is sampled at time $t_0, t_1, \ldots, t_{k+1}$, $r$ is a time-varying function describing the velocity of spread and $r_{k+1}$ is the velocity of spread $r$ for $(k+1)$th sample.

The state estimates are updated using (4.10). The measurement equations are given by (4.11) and (4.12) with noise covariance $v_1 \sim N(0, R_1)$. The time and measurement update equations are given by

$$\begin{aligned} &\hat{I}^-_{k+1} = \hat{I}_k, \qquad \hat{\sigma}^-_{k+1} = \hat{\sigma}_k, \qquad (\hat{y}_0)^-_{k+1} = (\hat{y}_0)k \\ &(\hat{x}_0)^-_{k+1} = (\hat{x}_0)_k + r_{k+1}\Delta t_{k+1} \\ &P^-_{k+1} = P_k + Q_{k+1} \end{aligned} \tag{4.11}$$

and

$$\begin{aligned} \hat{H}_{k+1} &= \frac{\partial \hat{g}_{k+1}}{\partial(\hat{A}^-_{k+1})} = \left[\frac{\partial g}{\partial \hat{I}^-_0} \quad \frac{\partial g}{\partial \hat{\sigma}^-} \quad \frac{\partial g}{\partial \hat{x}^-_0} \quad \frac{\partial g}{\partial \hat{y}^-_0}\right]_{k+1} \\ &= \left[C \quad \hat{I}^-_{k+1}(\hat{\sigma}^-_{k+1})^3\{(x_{k+1} - \hat{x}^-_{0k+1})^2 + (y_{k+1} - \hat{y}^-_{0k+1})^2\}C \right. \\ &\qquad \left. \times \frac{\hat{I}^-_{k+1}(x_{k+1} - \hat{x}^-_{0k+1})}{(\hat{\sigma}^-_{k+1})^2}C \quad \frac{\hat{I}^-_{k+1}(y_{k+1} - \hat{y}^-_{0k+1})}{(\hat{\sigma}^-_{k+1})^2}C\right] \end{aligned}$$

where

$$C = \exp\left[-\frac{(x_{k+1} - \hat{x}^-_{0k+1})^2 + (y_{k+1} - \hat{y}^-_{0k+1})^2}{2(\hat{\sigma}^-_{k+1})^2}\right] \tag{4.12}$$

$$K_{k+1} = P^-_{k+1}\hat{H}^{\mathrm{T}}_{k+1}(\hat{H}_{k+1}P^-_{k+1}\hat{H}^{\mathrm{T}}_{k+1} + R_{k+1})^{-1} \quad P_{k+1} = (I_{12} - K_{k+1}\hat{H}_{k+1})P^-_{k+1}$$

$$\begin{aligned} \hat{A}_{k+1} &= \hat{A}^-_{k+1} + K_{k+1}\{\tilde{z}_{k+1} - \hat{g}_{k+1}\} \\ &= \hat{A}^-_{k+1} + K_{k+1}\left\{\tilde{z}_{k+1} - \hat{I}^-_{k+1}\exp\left[-\frac{(x_{k+1} - \hat{x}^-_{0k+1})^2 + (y_{k+1} - \hat{y}^-_{0k+1})^2}{2(\hat{\sigma}^-_{k+1})^2}\right]\right\} \end{aligned}$$

### *4.3.2 Formulation of the general multi-scale algorithm EKF-NN-GAS for fire fields*

In the previous sections, we considered different cases of increasing complexity, in which robots sample a parameterized field distribution. In a practical scenario, measurements will be arriving from different sensors, at different dimensions and timescales, and must all be fused into the EKF filter. We propose the following sampling algorithm to negotiate the increased complexity of mapping the fire field:

*Algorithm EKF-NN-greedy adaptive sampling (GAS) (multi-scale, multi-rate AS)*

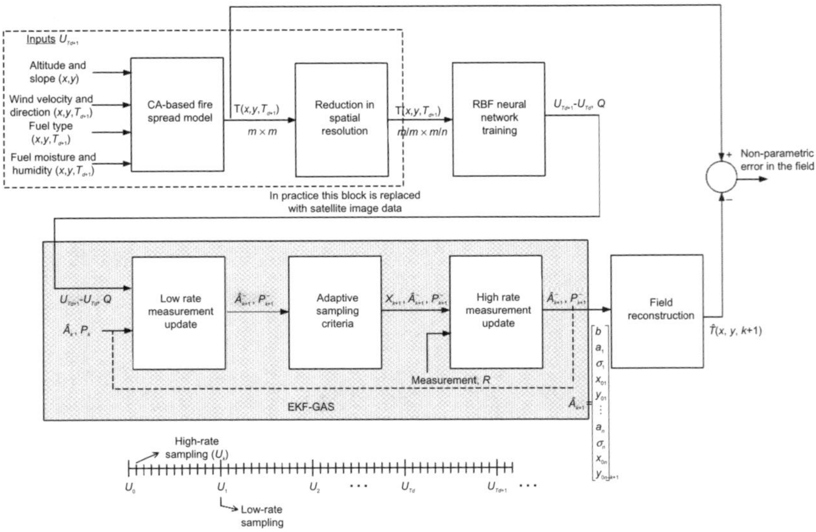


*Figure 4.8 Block diagram for temporal field model identification and parameter estimation*

*Step 1 (Setup)*: Give the environmental parameters and rules for fire spread as input to the CA model as shown in Figure 4.8. This will generate a 2D temperature field $T(x, y, t)$ that has a dimension of $m \times m$ at time $t$. In practice, this step is omitted and replaced by the actual spread of the fire. Go to Step 2.

*Step 2 (Initialization)*: Divide the field into square size grid of $n \times n$, $n < m$ and average values in each grid. This gives a low-resolution version of the actual field of size $m/n \times m/n$, illustrated in Figure 4.8. In practice, this low-resolution temperature distribution is acquired by an infrared image taken from an airborne platform. Go to Step 3.

*Step 3 (Training)*: Train the RBF neural network using this low-resolution temperature data. In the training algorithm, the number of neurons and smoothness factor are specified. The number of neurons depends on the complexity of the field so that the error is minimized within an acceptable threshold. This gives the parameterized version of the field with $N$ neurons, and each neuron has parameters $a$, $\sigma$, $x_0$ and $y_0$ representing this RBF field. The error in the actual field and the initial estimate using neural network also gives a guess for initial error covariance $P$ in later EKF steps. Go to Step 4.

*Step 4 (High rate sampling)*: Spot measurement robots sample locations in a grid of size $p \times p$ (where $p \leq n <$ m), based on a GAS criterion to minimize the error covariance. The EKF framework shown in Figure 4.8 is used to correct the estimates as the subsequent measurements are available one by one. The robot location is calculated by GPS measurement, via dead reckoning, or relative position measurements. Localization uncertainty is ignored in the simulation results in the next section but should be considered along with the robot dynamic model. The EKF sampling rate, $T$, should be as fast as sensory measurements from robots are available. Repeat Step 4 until new low-resolution updates of the entire field are available, else go to Step 5.

*Step 5 (Low rate sampling)*: Because of the time-varying nature of the field, a low-resolution update of the field is performed at a sampling rate $T_d$, and the evolution of the parameters is updated by comparing the low-resolution field distribution at time $T_{d+1}$ and $T_d$. This involves repeating Step 3 by retraining the neural network at time $T_{d+1}$ and updating the parameters since the last training at time $T_d$. The low rate parameter update $(\tilde{U}_{Td+1} - \tilde{U}_{Td})$ and its uncertainty $Q$ are given as inputs to the EKF block for high rate sampling update shown in Figure 4.8. Low rate sampling is performed at a very low rate (approximately every 5 min for the simulation examples we considered), compared with the high rate of sampling (which takes several sensor measurements every minute for the simulation examples). This scheme gives a better estimate of the time-varying field than the case when it was not included in the parameter time update. Go to Step 4.

Since the sampling robot dynamics is ignored here for the sake of simplicity, the parameter estimation model is simply

$$\begin{aligned} A_{k+1} &= A_k + \alpha_k \\ &= [b \;\; a_1 \;\; \sigma_1 \;\; x_{01} \;\; y_{01} \;\; \ldots \;\; a_N \;\; \sigma_N \;\; x_{0N} \;\; y_{0N}]_k^{\mathrm{T}} + \alpha, \quad \alpha \sim N(0, Q) \end{aligned} \tag{4.13}$$

Moreover, if the field is time-varying, the field parameters propagate as

$$A_{k+1} = A_k + (\tilde{U}_{Td+1} - \tilde{U}_{Td}) + \alpha \tag{4.14}$$

where $A_{k+1}$ is the field parameter update when $(k + 1)$th sample is taken (Step 4) and $(\tilde{U}_{Td+1} - \tilde{U}_{Td})$ is the field parameter propagation obtained by performing Step 5. The measurement model is

$$\begin{aligned} Z_k &= h(A_k) + v_k \\ &= b + \sum_{i=1}^{N} a_i \exp[-\sigma_i\{(x_k - x_{0i})^2 + (y_k - y_{0i})^2\}] + v, \quad v \sim N(0, R) \end{aligned} \tag{4.15}$$

while the low rate sampling update equations are given by

$$\begin{aligned} \hat{A}_{k+1}^{-} &= \hat{A}_k + (\tilde{U}_{Td+1} - \tilde{U}_{Td}) \\ P_{k+1}^{-} &= P_k + Q \end{aligned} \tag{4.16}$$

Step 5 does not reset the previous state estimates but calculates the evolution of parameters $(\tilde{U}_{Td+1} - \tilde{U}_{Td})$ based on the difference between current and past field images added to the old parameter estimate $\hat{A}_k$ to predict a new estimate $\hat{A}_{k+1}^{-}$.

Finally, equations similar to (4.12) can be used in the EKF sampling update. Figure 4.8 shows a block diagram for algorithm *EKF-NN-GAS*. Its full update equations are given by the following block equations:

$$
\begin{aligned}
\hat{H}_{k+1} &= \frac{\partial \hat{h}_{k+1}}{\partial (\hat{A}_{k+1}^{-})} \\
&= \left[ \frac{\partial \hat{h}_{k+1}}{\partial \hat{b}_{k+1}^{-}} \; \frac{\partial \hat{h}_{k+1}}{\partial (\hat{a}_1)_{k+1}^{-}} \; \frac{\partial \hat{h}_{k+1}}{\partial (\hat{\sigma}_1)_{k+1}^{-}} \; \frac{\partial \hat{h}_{k+1}}{\partial (\hat{x}_{01})_{k+1}^{-}} \; \frac{\partial \hat{h}_{k+1}}{\partial (\hat{y}_{01})_{k+1}^{-}} \right. \\
&\qquad \left. \cdots \; \frac{\partial \hat{h}_{k+1}}{\partial (\hat{a}_N)_{k+1}^{-}} \; \frac{\partial \hat{h}_{k+1}}{\partial (\hat{\sigma}_N)_{k+1}^{-}} \; \frac{\partial \hat{h}_{k+1}}{\partial (\hat{x}_{0N})_{k+1}^{-}} \; \frac{\partial \hat{h}_{k+1}}{\partial (\hat{y}_{0N})_{k+1}^{-}} \right] \\
&= [\, 1 (\hat{C}_1)_{k+1} - 2(\hat{a}_1)_{k+1}^{-} (\hat{\sigma}_1)_{k+1}^{-} (\hat{C}_1)_{k+1} (\hat{D}_1)_{k+1} \\
&\qquad 2(\hat{a}_1)_{k+1}^{-} (\hat{\sigma}_1)_{k+1}^{-} (\hat{C}_1)_{k+1} \{x_{k+1} - (\hat{x}_{01})_{k+1}^{-}\} \\
&\qquad 2(\hat{a}_1)_{k+1}^{-} (\hat{\sigma}_1)_{k+1}^{-} (\hat{C}_1)_{k+1} \{y_{k+1} - (\hat{y}_{01})_{k+1}^{-}\} \\
&\qquad \cdots \; (\hat{C}_N)_{k+1} - 2(\hat{a}_N)_{k+1}^{-} (\hat{\sigma}_N)_{k+1}^{-} (\hat{C}_N)_{k+1} (\hat{D}_N)_{k+1} \\
&\qquad 2(\hat{a}_N)_{k+1}^{-} (\hat{\sigma}_N)_{k+1}^{-} (\hat{C}_N)_{k+1} \{x_{k+1} - (\hat{x}_{0N})_{k+1}^{-}\} \\
&\qquad 2(\hat{a}_N)_{k+1}^{-} (\hat{\sigma}_N)_{k+1}^{-} (\hat{C}_N)_{k+1} \{y_{k+1} - (\hat{y}_{0N})_{k+1}^{-}\}
\end{aligned} \tag{4.17}
$$

where

$$
\begin{aligned}
(\hat{C}_i)_{k+1} &= \exp\{-((\hat{\sigma}_i)_{k+1}^{-})^2 (\hat{D}_i)_{k+1}\} \\
(\hat{D}_i)_{k+1} &= \{(x_{k+1} - (\hat{x}_{0i})_{k+1}^{-})^2 + (y_{k+1} - (\hat{y}_{0i})_{k+1}^{-})^2\} \\
K_{k+1} &= P_{k+1}^{-} \hat{H}_{k+1}^{\mathrm{T}} (\hat{H}_{k+1} P_{k+1}^{-} \hat{H}_{k+1}^{\mathrm{T}} + R_{k+1})^{-1} \\
P_{k+1} &= (I - K_{k+1} \hat{H}_{k+1}) P_{k+1}^{-} \\
\hat{A}_{k+1} &= \hat{A}_{k+1}^{-} + K_{k+1} \left\{ \tilde{z}_{k+1} - \hat{h}_{k+1} \right\} \\
&= \hat{A}_{k+1}^{-} + K_{k+1} \{ \tilde{z}_{k+1} - \hat{b}_{k+1}^{-} - \sum_{i=1}^{N} (\hat{a}_i)_{k+1}^{-} \exp[-((\hat{\sigma}_i)_{k+1}^{-})^2 \\
&\qquad \times \{(x_{k+1} - (\hat{x}_{0i})_{k+1}^{-})^2 + (y_{k+1} - (\hat{y}_{0i})_{k+1}^{-})^2\}] \Big\}
\end{aligned}
$$

## 4.4 Potential field to aid navigation through fire field using mobile agents

The estimated fire field intensity distribution can be used as a repulsive potential to keep the firefighters away from dangerous areas in the field and show them safe paths towards important destinations. Potential field methods create a vector field

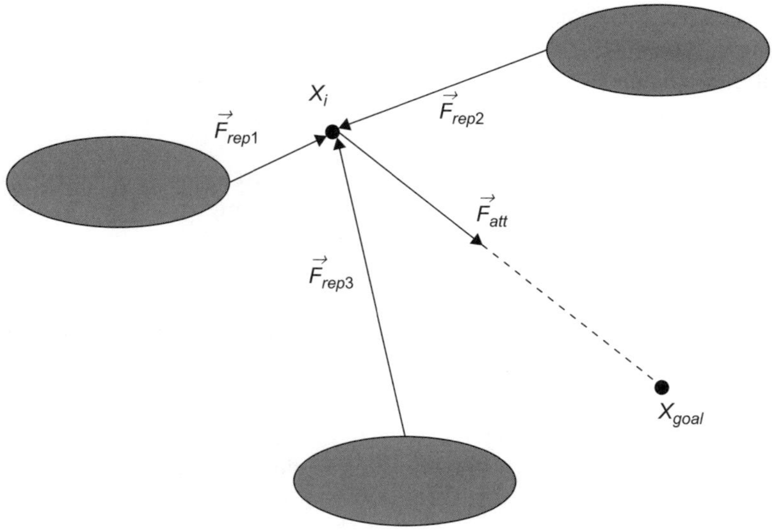


*Figure 4.9 The attractive forces on point $X_i$ towards the goal $X_{goal}$ and repulsive forces from the obstacles*

representing a navigational path based on a potential function. The sampled field variable intensity can be used to plan collision-free paths around the fire 'obstacles'. Given a scalar potential function $U(X)$, where $X = (x, y)$, which depends on the rescuer position and the field intensity at that point, one can calculate forces governing the rescuer motion based on the gradient of the scalar potential field (Figure 4.9):

$$\vec{F}(X) = -\vec{\nabla}U(X) \tag{4.18}$$

The following forces can be considered in the potential field:

- Attractive forces towards goals:

$$\vec{F}_{att}(X) = -\vec{\nabla}U_{att}(X) = -\xi\rho_{goal}(X)\vec{\nabla}\rho_{goal}(X) = \xi(X - X_{goal}) \tag{4.19}$$

- Repulsive forces from obstacles that are fire ellipses:

$$U_{rep}(X) = \sum_{k=1}^{n} U_k(X), \qquad U_k(X) = \lambda\frac{I_k}{\rho_k(X)^2} \tag{4.20}$$

$$\vec{F}_{rep}(X) = \sum_{k=1}^{n} \vec{F}_k(X), \qquad \vec{F}_k(X) = -\vec{\nabla}U_k(X) \tag{4.21}$$

where $\lambda$ is the positive scaling factor, $\rho(X)$ is the Euclidean distance from $X$ to the centre of ellipse, $n$ is the number of elliptical components close to current location and $I_k$ is the fire intensity at the point on ellipse at shortest distance from $X$. The trajectory can then be updated using a depth-first planning algorithm, which constructs a path as the product of successive segments starting at the initial configuration $X_i$:

$$\vec{X}_{i+1} = \vec{X}_i + \delta_i\vec{F}_p \tag{4.22}$$

where $X_i$ and $X_{i+1}$ are the origin and end extremities of the $i$th segment in the path.

In summary, this chapter discusses the application of the AS algorithm to the mapping of forest fires. In addition to the basic approach discussed in Chapter 3, we also offer a complete framework encapsulated in the algorithm EKF-NN-GAS, which combines sensor measurements from different scales, rates and accuracies to map the time-varying spread of forest fires.

Finally, we need to briefly discuss issues related to the convergence of the proposed algorithms. Since we are using the EKF, any estimation scheme that utilizes it must be initialized sufficiently close to the actual field. Overhead satellite imagery of the field provides a reasonable initial estimate, but the absolute algorithm convergence cannot be guaranteed, as widely discussed in the literature [101]. In addition, the use of heuristic search methods in the algorithm can also lead to the presence of local minima. We can avoid such minima by restricting the search space so that we do not revisit already sampled points, but such heuristics may not always work for time-varying fields. While studying convergence conditions for our algorithms is beyond the scope of this book, simulation results presented in the next section indicate that the algorithms converge in numerous instances.

## 4.5 Simulation results

In this section, increasingly realistic fire spread models are used to assess the effectiveness of the sampling algorithm by comparing it to a basic raster scanning sampling approach. Also, the performance of the multi-rate EKF scheme to estimate the time-varying fire field model is investigated by simulation. First, sampling results on a single Huygens' spread model are presented. We then consider the case of a field represented by a sum of five Gaussians with slow spreading over time, and finally, the case of a time-varying CA fire spread complex model. In all simulations, sampling is performed with a single robotic vehicle. For multiple vehicles, the EKF computations can be distributed among robots as discussed in Chapter 5. The model of the robots is ignored in order to focus on quantifying the accuracy of field estimation.

### *4.5.1 Elliptically constrained single Gaussian time-varying forest fire field*

Simulations are performed to estimate four fire field parameters given in (4.10)–(4.12) for a single ellipse. The numerical simulation model assumed a forest area of 1 square mile, in which multiple robots take fire intensity measurements (or some measure of temperature) and estimate the desired parameters. The location of firehead $x_0$ is time-varying. The fire can spread at arbitrary rates, but in this simulation, the case of a slow sinusoidal fire spread rate was considered as an example:

$$r = \frac{2\pi}{60}\cos\left(\frac{2\pi}{60}t\right) \tag{4.23}$$

Other choices of rate spread will not affect the rate of convergence as long as the spread model is known, and the rate of spread is much smaller than the speed of sampling. As a result of the fire spread rate in (4.23), $x_0$ will be sinusoidal with respect to time and space, while the other fire field parameters are stationary. A comparison sampling simulation was conducted between sampling using raster scanning and heuristic GAS. For GAS, the algorithm looks for the next best sampling location in a circle of 50-ft radius around the currently sampled location as shown in Figure 4.10. Raster scanning does a row-by-row scanning.

Table 4.1 summarizes the sampling results of GAS and raster scanning when sampling is performed for 60 min. We assumed that the robot navigation speed was

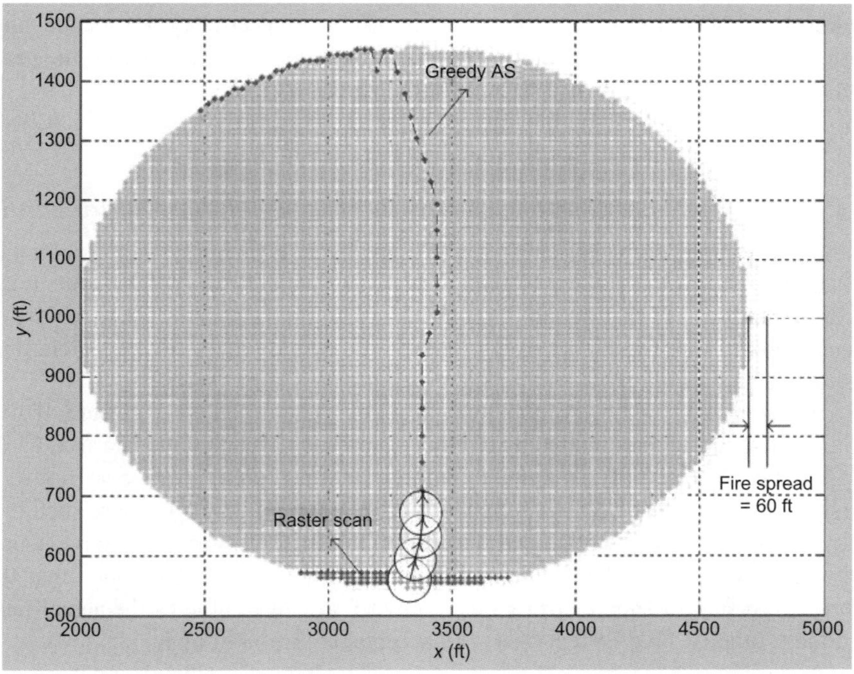


*Figure 4.10 Elliptical fire spread with greedy AS algorithm that looks for the appropriate location on the elliptical fire*

*Table 4.1 Comparison of raster scan and greedy AS for sampling of elliptically constrained single Gaussian time-varying field*

| | $A_0$ | $A_k$ | $\hat{A}_0$ | $\hat{A}_k$ | $P_0$ | $P_k$ |
|---|---|---|---|---|---|---|
| Greedy AS ($k = 49$) | | | | | | |
| $I$ | 80 | 80 | 90 | 79.4 | 10 | 0.7 |
| $\sigma$ | 400 | 400 | 410 | 401.1 | 20 | 4.9 |
| $x_0$ | 2700 | 2760 | 2720 | 2759.7 | 50 | 3.8 |
| $y_0$ | 1000 | 1000 | 1020 | 1001.2 | 50 | 3.8 |
| Raster scan ($k = 68$) | | | | | | |
| $I$ | 80 | 80 | 90 | 85.7 | 10 | 2.4 |
| $\sigma$ | 400 | 400 | 410 | 397.9 | 20 | 5.1 |
| $x_0$ | 2700 | 2760 | 2720 | 2765.5 | 50 | 21.8 |
| $y_0$ | 1000 | 1000 | 1020 | 1024.8 | 50 | 38.1 |

30 ft/min, and the robot sampling and processing time are neglected, while $Q = 0$ and $R_1 = 0.1$.

Results indicate that GAS leads to faster convergence and requires considerably less number of sampling points than RS or AS. In Table 4.1, the 2-norm of the error covariance $P_k$ is 7.3 for GAS and 44.2 for raster scanning after 60 min of sampling. If sampling continues further, raster scanning takes almost six times more time than GAS to converge to same parameter estimate values. We also observed that the error covariance decreases very slowly for raster scanning, and it will require almost the entire sampling area to reduce the parameter estimate uncertainty. Raster scanning performs even worse when sampling is being conducted in an area where the parameter of interest does not vary significantly (Figure 4.11).

### 4.5.2 RBF-NN parameterization using low-resolution information

Radial basis neural network using 'newrb' is available in neural network toolbox of MATLAB. 'newrb' creates a two-layer network, with the first layer containing 'radbas' neurons and the second, 'purelin' neurons. In this section, we make use of this toolbox to simulate the sampling process of a slightly more complex field than a single Gaussian.

In Figure 4.12a (see p. C10), a simple field is presented where a sum of five Gaussians field is approximated by an RBF neural network with five neurons, while in Figure 4.12b (see p. C10) a more complex field is displayed. A low-resolution version of the original fields is acquired by averaging points in a square such that only a small percentage of the total number of points is used for training the neural network. The number of neurons and the spread factor are chosen such that the 'normalized sum-of-square error (SSE)' between low resolution of the actual field and the estimated field is kept below an acceptable threshold of 1.

As discussed in Chapter 1, the neural network learning algorithm does not train for spread parameters. Although some heuristic algorithms exist to select different values for spread factor, we are assuming a constant spread value. For our simulations, $p = 300$, a typical case where grid size is $n \times n = 20 \times 20$ and spread factor is 30 with five neurons is assumed as shown in Figure 4.12a. The 2-norm of error between actual and initial estimated fields is 38.7, and error extrema are $-144$ and 204.

A similar scheme is used to approximate the complex field, which is generated using CA. Since the field is more complex, more neurons are required for a good approximation. At a particular time, a low-resolution version of field is taken by averaging on a grid size of $20 \times 20$ and passed through RBF-NN training for 20 neurons and spread factor of 40. The training using the hybrid algorithm only takes 2–3 s for a 60-neuron neural network. Hence, the training time is much smaller than the speed of field evolution.

Figure 4.13 (see p. C11) also illustrates the use of self-organized selection of centres classification algorithm used for selection of centres while training the neural network. The objective of the classification is to introduce more neurons in high-variance areas of the field, and fewer neurons in low-variance areas. Figure 4.13 shows a sum of 30 Gaussians approximated with a 30-neuron RBF neural network.

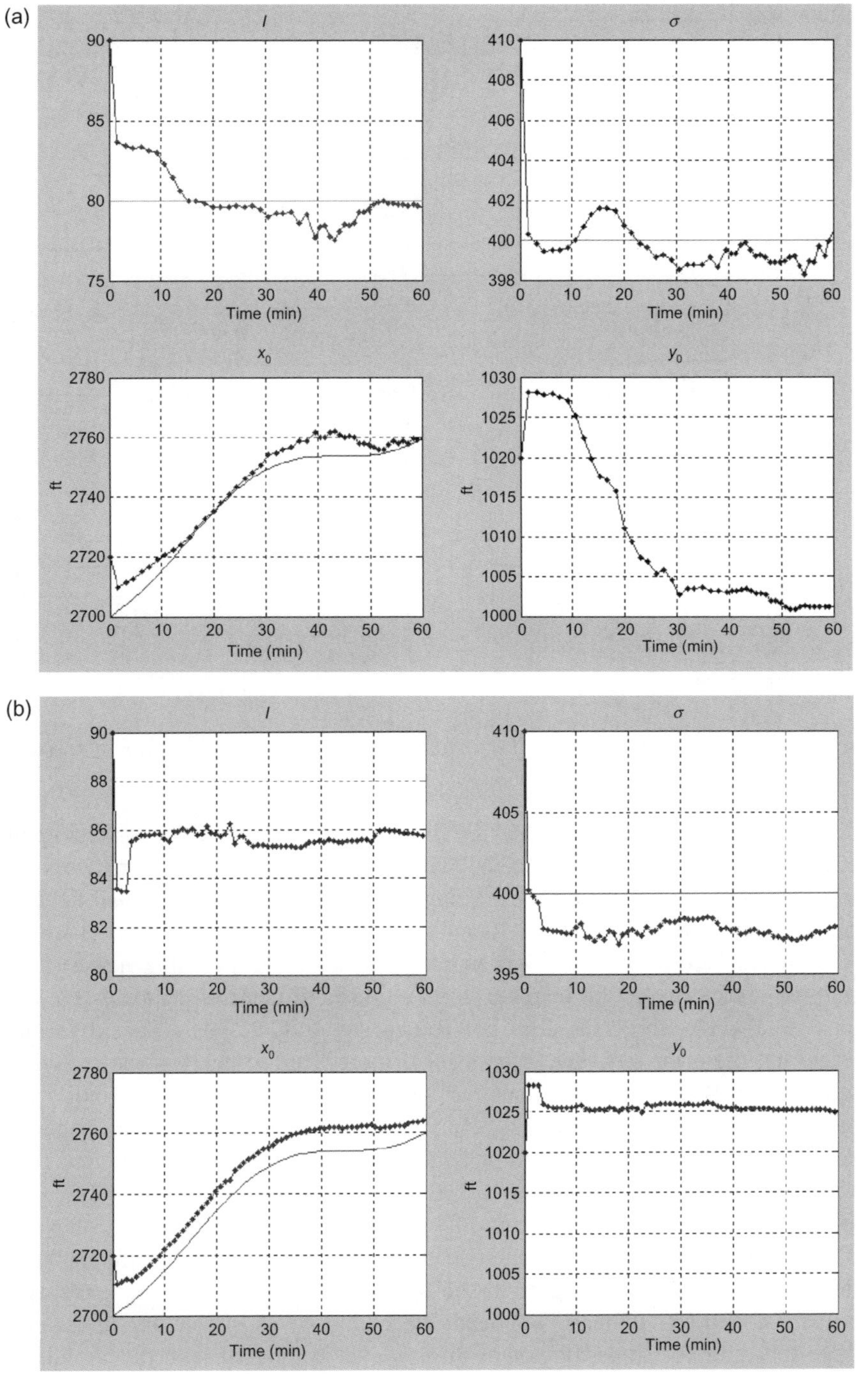


*Figure 4.11 Actual (solid lines) and estimated (dotted lines) fire field parameters I, σ, $x_0$ and $y_0$ versus time when sampling operation is performed for 60 min using (a) greedy approach and (b) raster scanning*

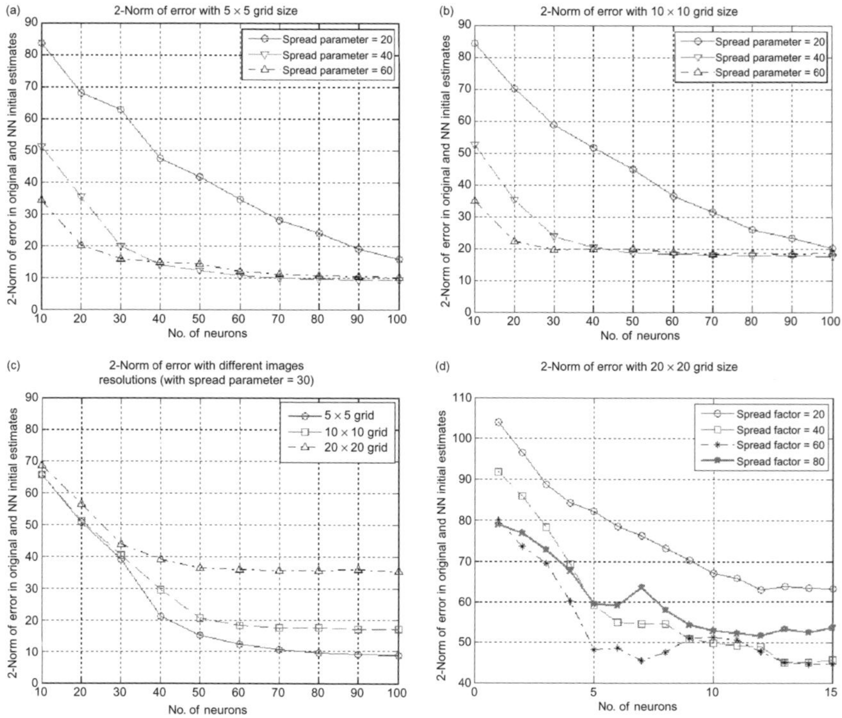


*Figure 4.14 Effect of number of neurons, spread factor and number of training points on the error in estimate*

The plots shown in Figure 4.14 are the 2-norm of relative error of all the points between the actual fire field and the field estimated by the neural network. The estimated field is achieved by considering lower resolutions of the actual field with grid sizes of $n \times n = 5 \times 5$, $10 \times 10$ and $20 \times 20$. In Figure 4.14a, a $5 \times 5$ size grid is considered; hence, the error is smaller compared with Figure 4.14b, where a $10 \times 10$ size grid is considered. An increase in the number of neurons decreases the error, but after a while, the error does not reduce any further. For a spread factor of 40 and 60, the error stays the same even if more than 40 neurons are considered. Figure 4.14c illustrates the obvious fact that taking smaller size grid (indicating a higher resolution) increases the accuracy of initial estimate with same number of neurons. As the number of neurons increases, the initial estimate gets better. This is valid until the neural network becomes overtrained. Figure 4.14d shows the error in approximating a sum-of-five Gaussians field with RBF-NN with different number of neurons when a $20 \times 20$ size grid and different spread factors are considered. Increase in the spread factor decreases the error as the number of neurons increase but leads to saturation as shown in Figure 4.14d where the error for spread factor of 80 is higher compared with 60. Simulations are performed to estimate the 21 parameters of the field represented by a sum of five Gaussians. The system is parameterized using neural network (NN) as discussed in Chapter 1, and the sampling results are shown in Figure 4.14.

In our simulations, additional numerical assumptions for uncertainties were as follows:

$$P_{b0} = 200, \quad P_{a0} = 50, \quad P_{\sigma 0} = 10^{-8}, \quad P_{x00} = 4, \quad P_{y00} = 4$$
$$Q = 0, \quad R = 1 \tag{4.24}$$

Figure 4.15 (see p. C11) shows the NN approximate when an actual fire field satellite image was used to train the neural network. For a complex continuous field, the following heuristic value for the spread factor was used to improve the NN training error: $s = \sqrt{xy/N}$, where $s$ is the spread parameter, $x$ and $y$ are maximum values of coordinates, and $N$ is the number of neurons.

### 4.5.3 Sum-of-Gaussians stationary field

A comparison simulation of GAS and raster scan sampling was carried out for the sum of five Gaussians non-linear field. The sampling area was divided into square grids, and several search horizons were considered. For greedy sampling, a grid size of $n = 5, p = 5$ and horizon size of five square grids were assumed. The simulation stopped when the 2-norm error between the actual and estimated fields reduced below 15. It is depicted in Figures 4.16f and 4.17f that at the start of sampling represented by 'black' circles with centre and radius $(\hat{x}_{k+1}, \hat{y}_{k+1}, \hat{\sigma}_{k+1})$, coincide with the 'solid line' circles with centre and radius $(\hat{x}_0, \hat{y}_0, \hat{\sigma}_0)$, but start moving towards the 'dotted line' circles that have centre and radius $(x, y, \sigma)$, as the sampling continues. Simulation results for raster scan sampling are shown in Figure 4.16, where it took 170 samples to achieve the norm of error in the original and estimated fields less than 15, while the 2-norm of error covariance reduced from 229.5 to 10.72. Simulation results for GAS are shown in Figure 4.17 (see p. C12), where sampling required 41 points to achieve the norm of error in the original and estimated fields less than 15, and the 2-norm of error covariance to reduce from 229.5 to 2.04. A comparison between initial and final errors is shown in Figures 4.16 and 4.17h and i.

Table 4.2 clearly indicates the GAS performs much better than raster scan in terms of sampling distance (time) as well as number of samples. Also, it is apparent in Figures 4.16 and 4.17 that the norm of error between the original and estimated fields, as well as the norm of state error covariances, decreases faster in the case of GAS (Figures 4.18–4.22).

### 4.5.4 Sum-of-Gaussians time-varying field

Here we make the sum of five Gaussians to vary slowly, and we rerun our sampling algorithms. Depending on the anticipated variation of the field parameters based on the CA model and NN inverse model, the inputs $B_k u_k$ will vary, but this variation is available to the EKF estimator according to (4.16).

The numerical values for simulation uncertainties were as follows:

$$\begin{aligned} &P_{b0} = 200, \quad P_{a0} = 50, \quad P_{\sigma 0} = 10^{-8}, \quad P_{x00} = 4, \quad P_{y00} = 4 \\ &Q_{b0} = 0.5, \quad Q_{a0} = 3, \quad Q_{\sigma 0} = 0, \quad Q_{x00} = 0, \quad Q_{y00} = 0 \\ &R = 1 \end{aligned} \tag{4.25}$$

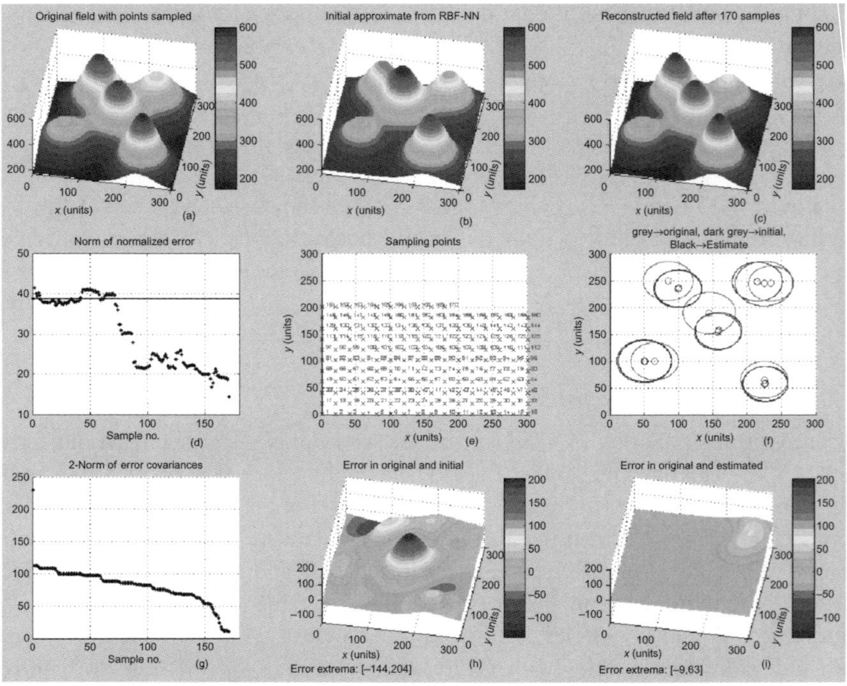


*Figure 4.16* *(a–i) Simulation results with raster scanning sampling for sum-of-Gaussians stationary field*

*Table 4.2* *Comparison of raster scan and greedy AS for sum of Gaussians field*

| | **Greedy AS** | **Raster scan** |
|---|---|---|
| Distance | 2175 | 3400 |
| No. of samples | 41 | 170 |
| Initial 2-norm of error in actual and estimated fields | 38.7 | 38.7 |
| Final 2-norm of error in actual and estimated fields | 14.97 | 14.26 |
| Initial 2-norm of error covariance | 229.5 | 229.5 |
| Final 2-norm of error covariance | 2.04 | 10.72 |

Figure 4.23 (see p. C12) shows the original field and the initial NN approximate by taking only 4 per cent samples from the original data. The error between the original and initial estimates is also shown in this figure. 1-$\sigma$ Gaussian circles of *initial* and *estimate* coincide with one another, and they are different from the *actual* Gaussian centres shown in blue colour.

Since the field is slow-varying, we assume that the time update is available slower than the rate at which measurements are taken (Figures 4.24–4.26) (see p. C13–14). In other words, robots take sensor measurements faster than the update

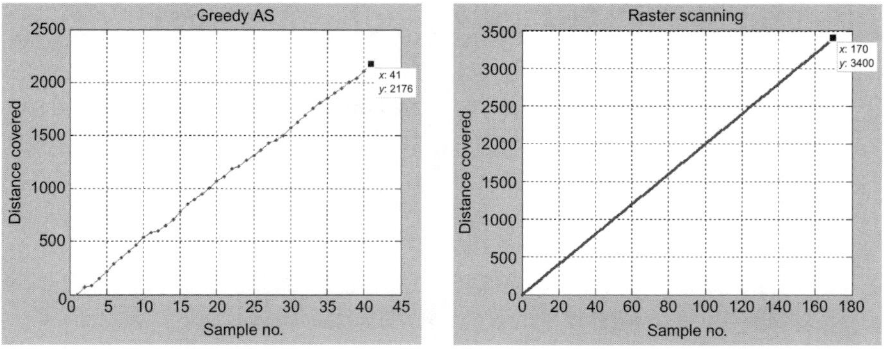


*Figure 4.18 Distance covered for greedy AS (left) and raster scanning (right)*

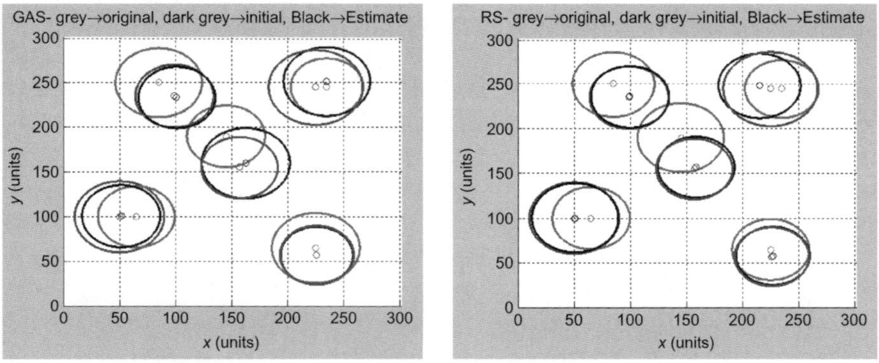


*Figure 4.19 Location of Gaussian centres initially and after sampling is done*

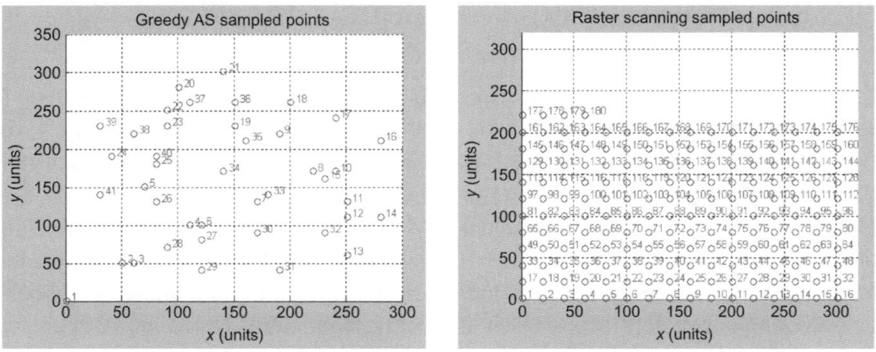


*Figure 4.20 Sampling points for GAS (left) and raster scanning (right)*

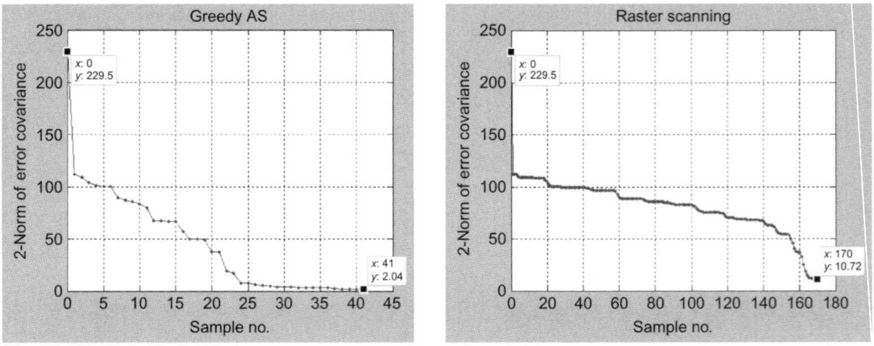


*Figure 4.21 2-Norm of error covariance for GAS (left) and raster scanning (right)*

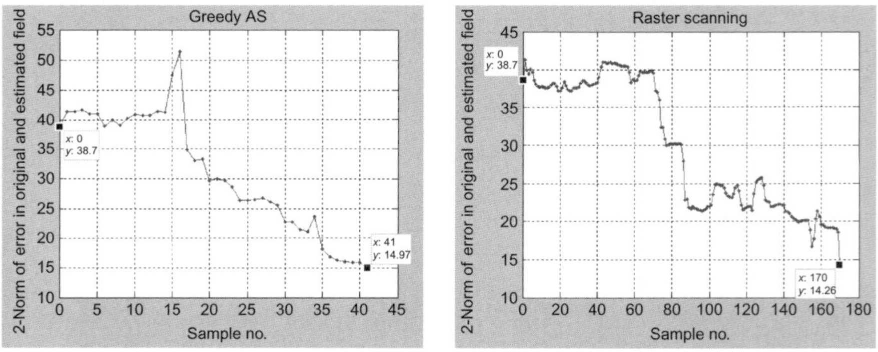


*Figure 4.22 2-Norm of error in original and estimated fields*

in parameters available from remote sensing measurement. When there is time update, the error covariance of the states goes higher but keeps on reducing again when measurements are taken. This observation is illustrated in Figure 4.27.

### 4.5.5 Complex RBF time-varying field

In this section, the complexity of the fire field model increases again by considering a slow time-varying field generated using the CA model presented in section 4.1.2. As the sampling algorithm uses an EKF, the observability of the parameters (algorithm convergence) will depend on the initial conditions. The initial error covariance is selected depending on the error in actual field and the initial estimated field, which in turn depends on the percentage of data from the actual field that is used for training the neural network, number of neurons and spread parameter. In our simulation models, the following parameters were chosen: the field is defined in an $m \times m = 300 \times 300$ area, and an average of values in an $n \times n = 30 \times 30$ grid is used for training the neural network. Forty neurons are used, with a spread parameter of 30. These parameters can

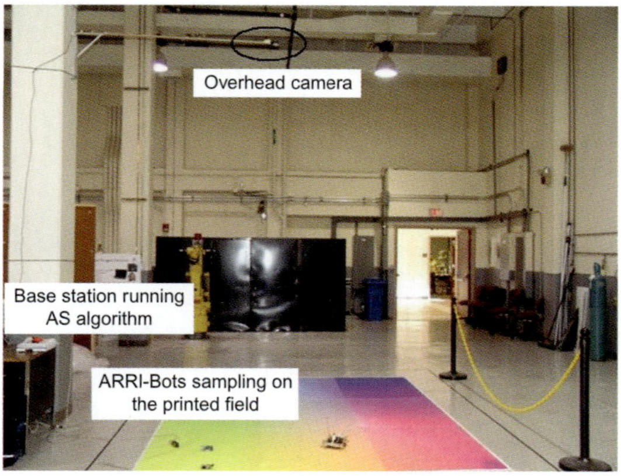


*Figure 2.2 The adaptive sampling test bed*

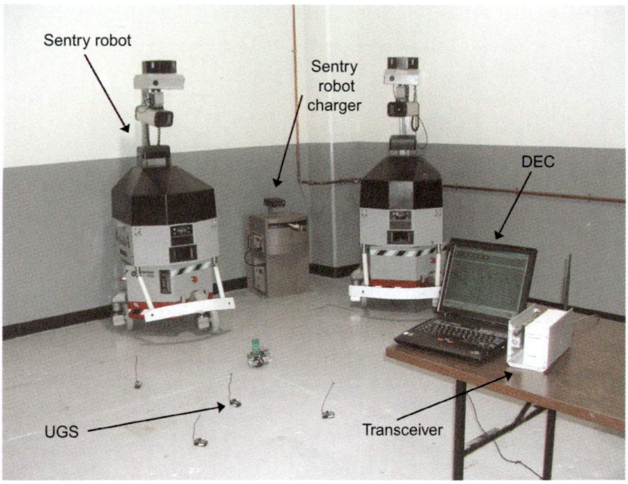


(a)

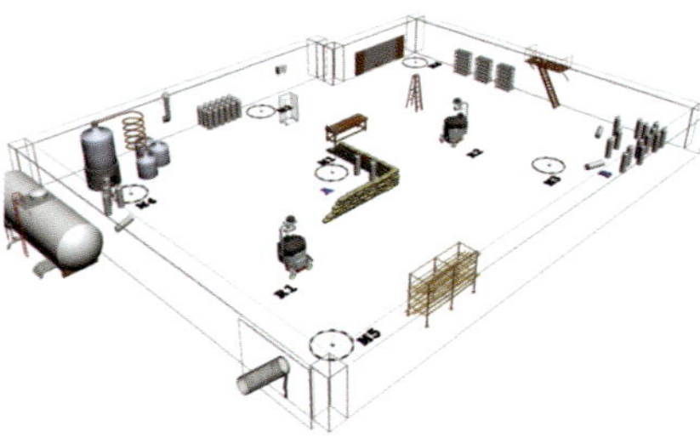

(b)

*Figure 2.10 The mobile WSN test bed at ARRI: (a) hardware and (b) virtual representation of the environment*

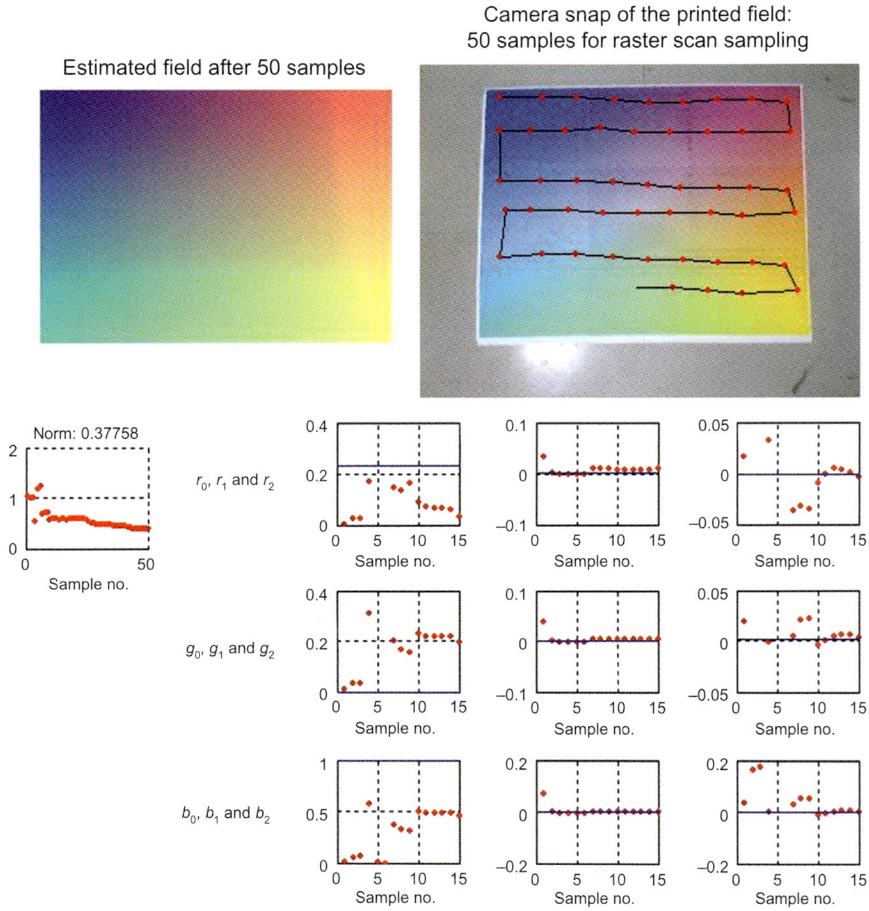


*Figure 3.9 Experimental results for raster scan sampling of linear colour field show estimated field after 50 samples. Field parameter estimates are plotted with respect to number of samples for the first 15 samples, and the 2-norm of the error is plotted with respect to the number of samples for 50 samples*

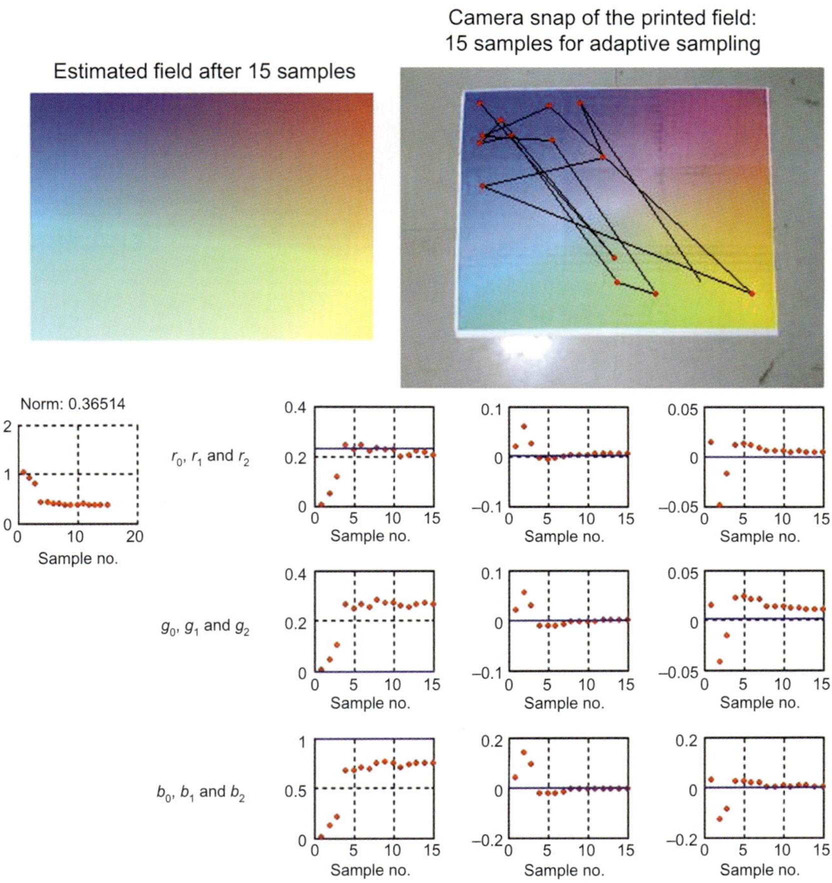


*Figure 3.10 Experimental results for AS show estimated field after 15 samples*

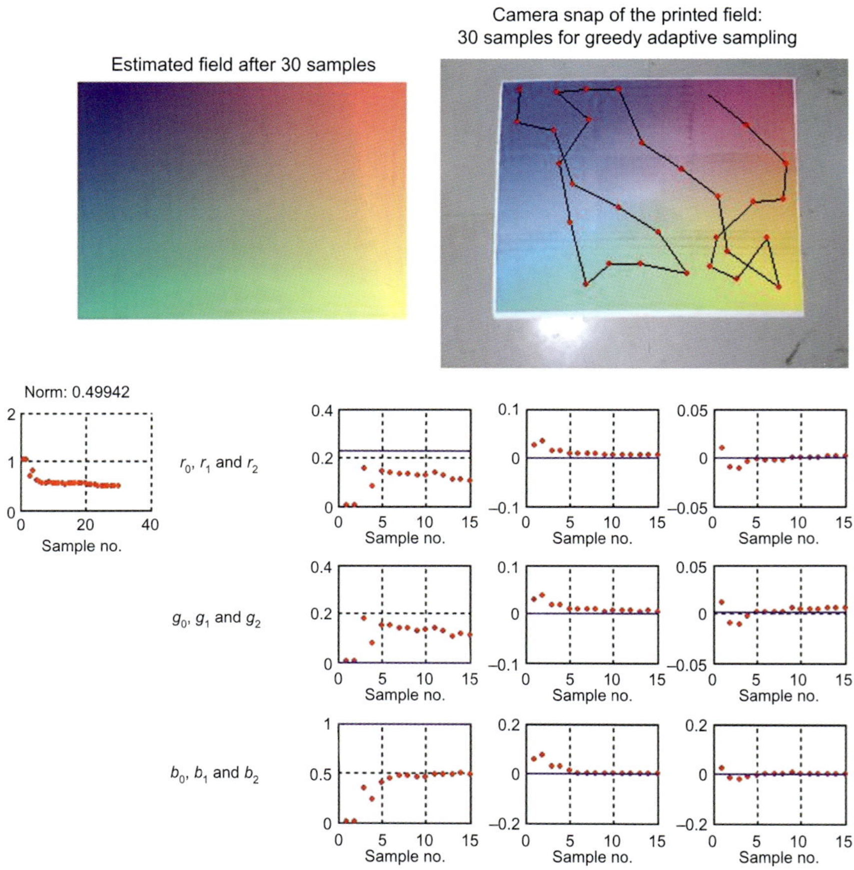


*Figure 3.11 Experimental results for GAS show estimated field after 30 samples. Field parameter estimates are plotted with respect to number of samples for the first 15 samples, and the 2-norm of the error is plotted with respect to the number of samples for 30 samples*

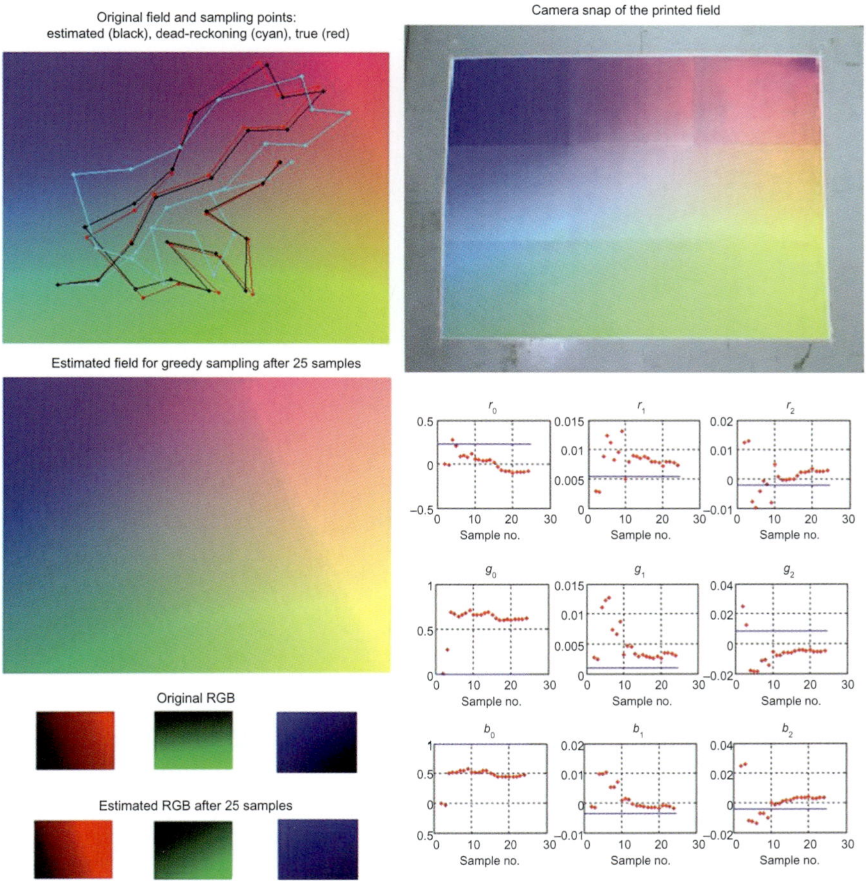


*Figure 3.16 Experimental results for GAS when location information from camera and colour measurement is fused in the EKF. Figure shows the field convergence after 25 samples considering the camera uncertainty*

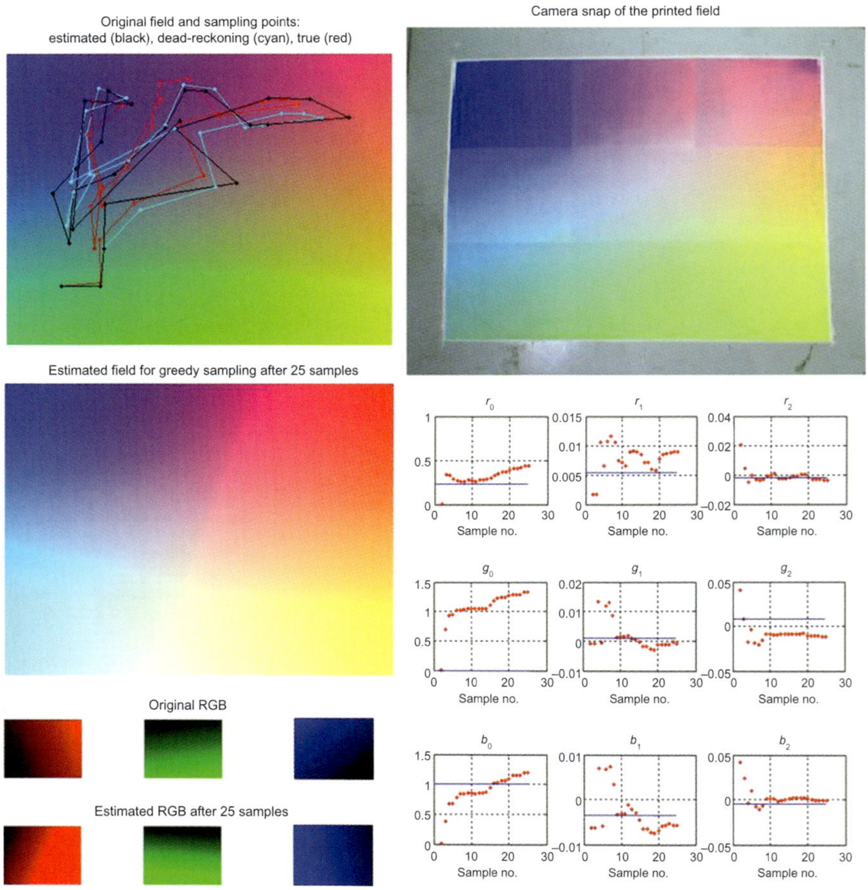


*Figure 3.17 Experimental results for GAS with location estimation from colour sensor measurement when location information from camera is not available. The figure shows the field convergence after 25 samples*

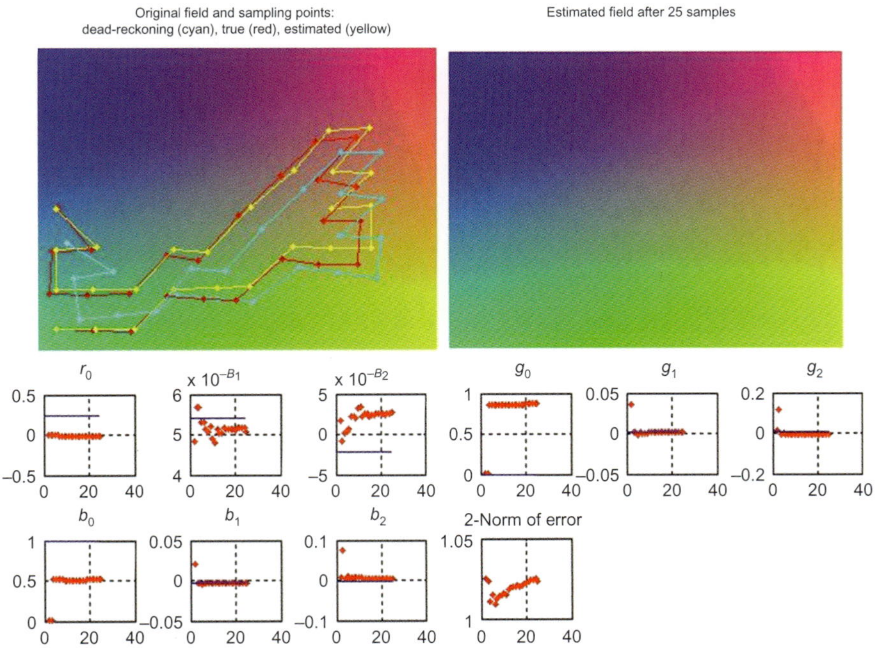


*Figure 3.18 Uncertainty of (0.01, 0.01, 0.01) in colour sensor measurement: field estimate converges and localization error is very small (after 25 samples)*

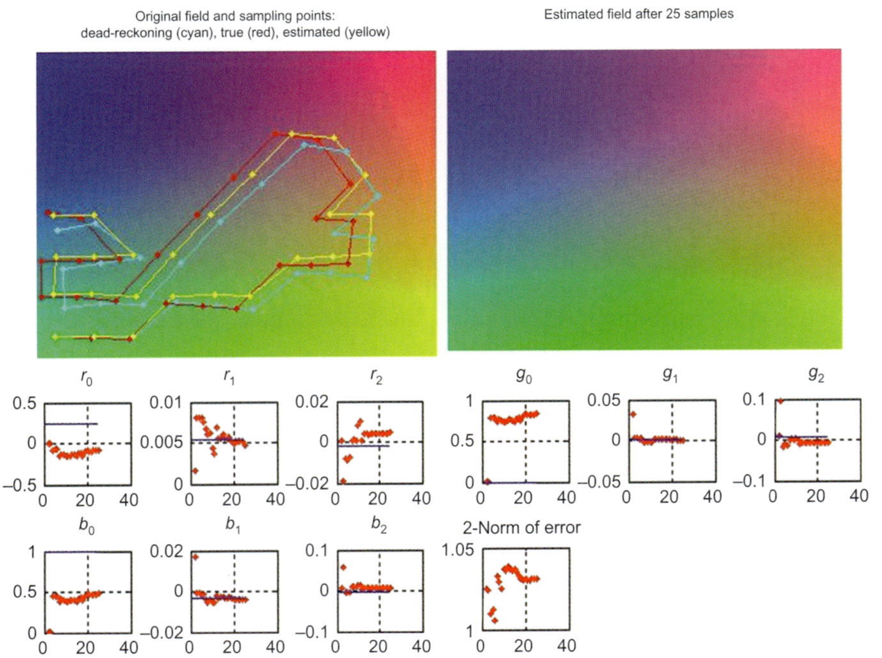


*Figure 3.19 Uncertainty of (0.1, 0.1, 0.1) in colour sensor measurement: field estimate converges and localization error is very small (after 25 samples)*

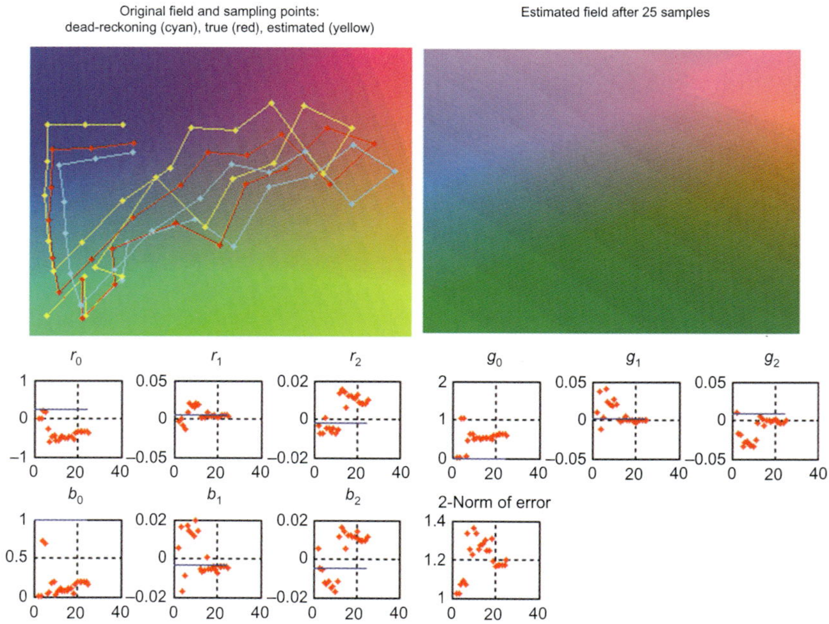


*Figure 3.20* *Uncertainty of (0.35, 0.35, 0.35) in colour sensor measurement: field estimate converges but with a marginal error and localization error is large*

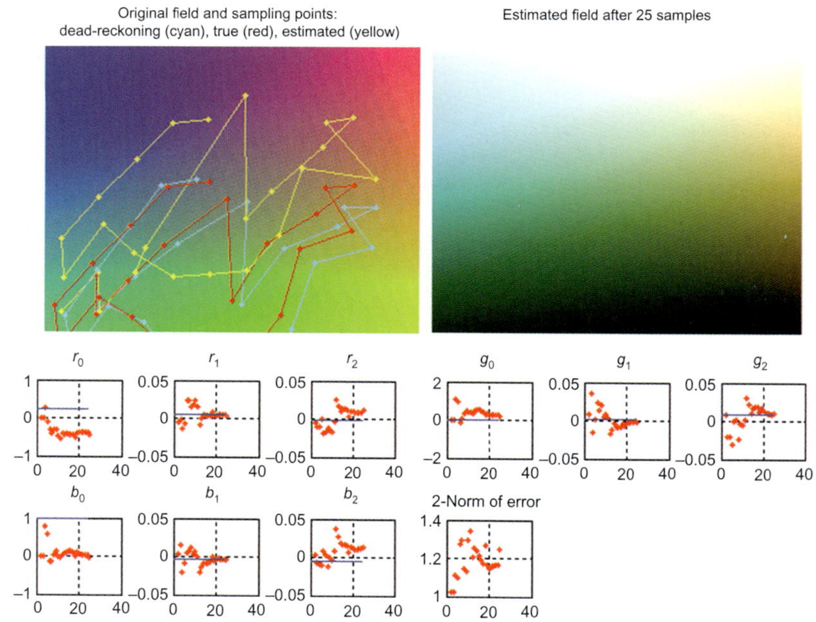


*Figure 3.21* *Uncertainty of (0.45, 0.45, 0.45) in colour sensor measurement: field estimate diverges and localization error is unacceptable (after 25 samples)*

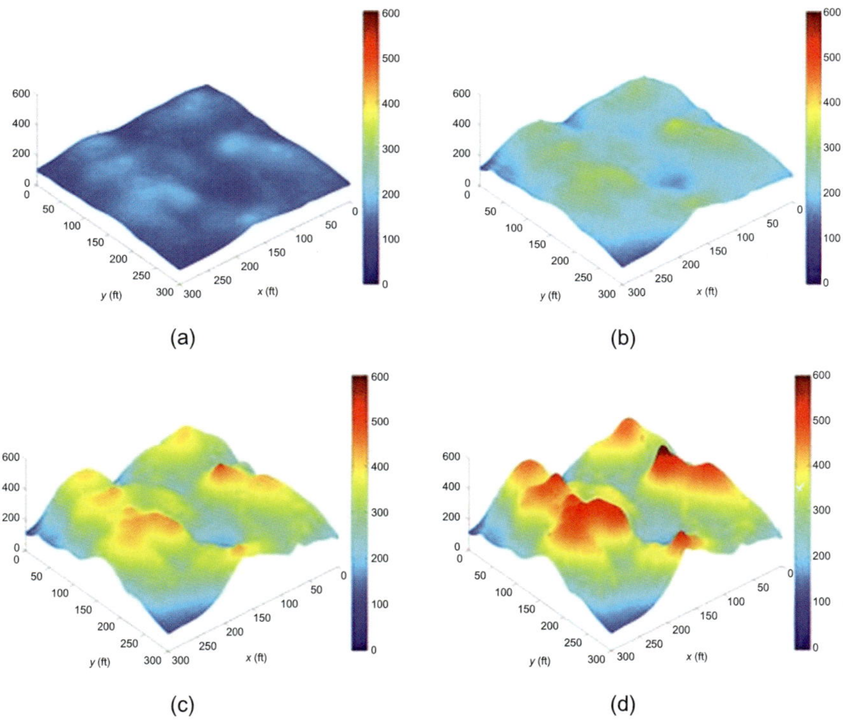


*Figure 4.6 Fire at time t =* 20, 40, 60 *and* 80 *min shown in (a), (b), (c) and (d), respectively*

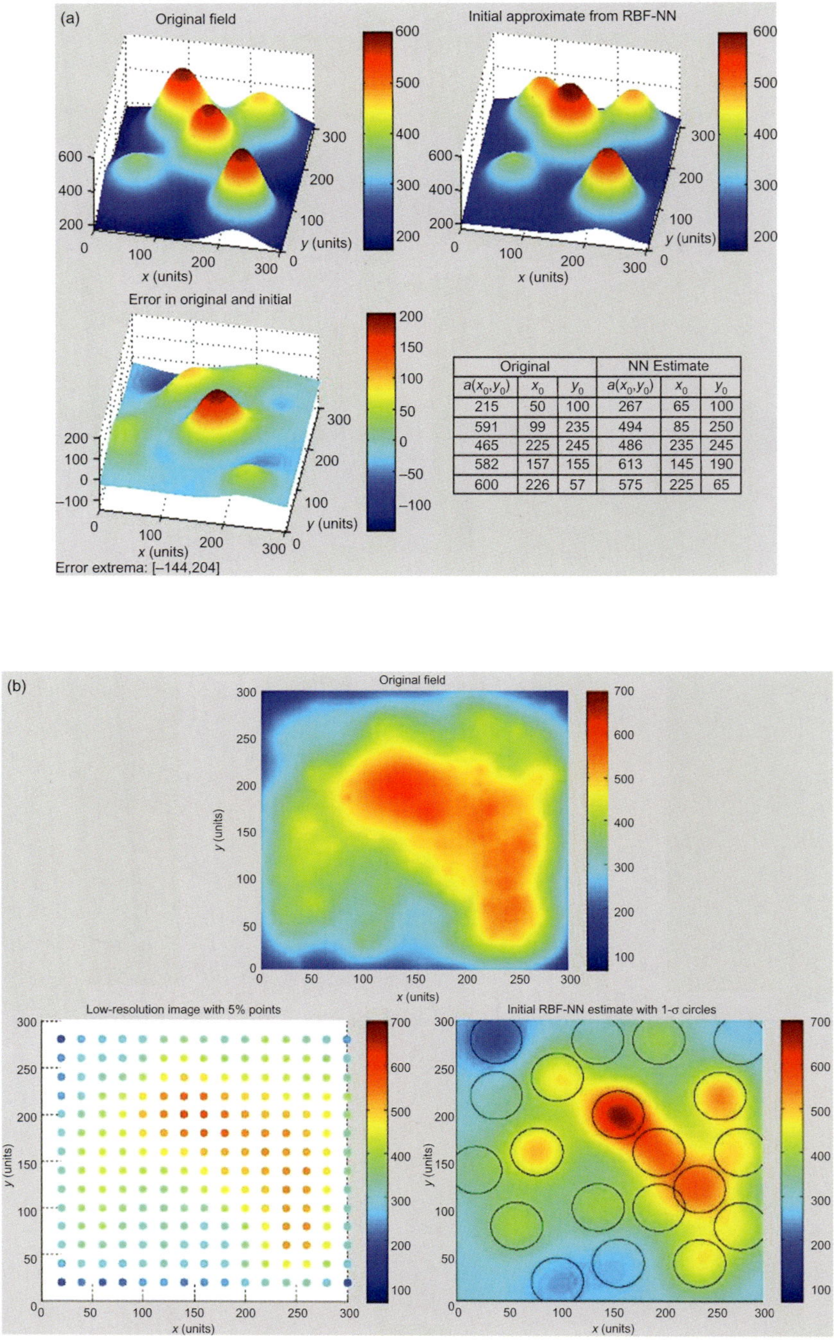


| Original | | | NN Estimate | | |
|---|---|---|---|---|---|
| $a(x_0,y_0)$ | $x_0$ | $y_0$ | $a(x_0,y_0)$ | $x_0$ | $y_0$ |
| 215 | 50 | 100 | 267 | 65 | 100 |
| 591 | 99 | 235 | 494 | 85 | 250 |
| 465 | 225 | 245 | 486 | 235 | 245 |
| 582 | 157 | 155 | 613 | 145 | 190 |
| 600 | 226 | 57 | 575 | 225 | 65 |

*Figure 4.12 (a) Sum-of-five Gaussian fire field approximation using 5 RBF neurons and (b) complex field approximation using 20 neurons and a grid size of 20 × 20*

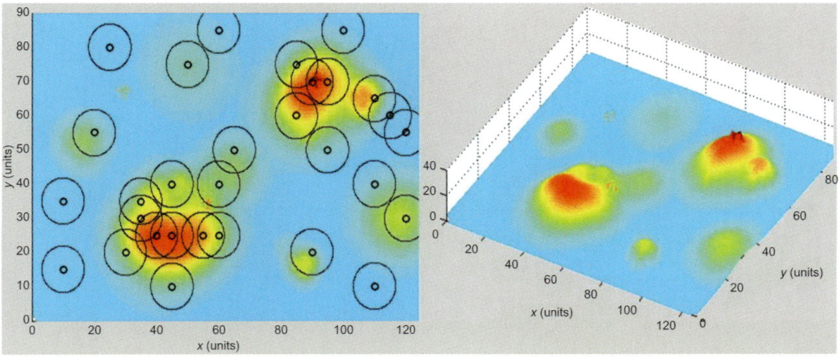


*Figure 4.13 Illustration of the presence of more overlapping Gaussians in areas of large variance for thorough sampling*

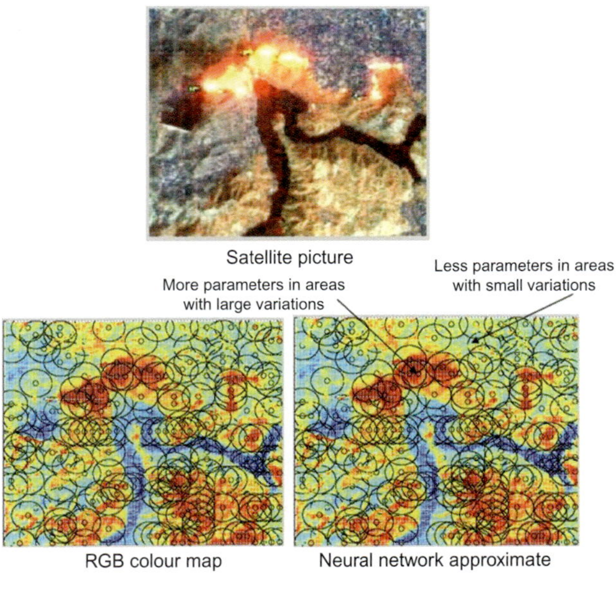


*Figure 4.15 Actual image of California 2007 fires taken from a NASA satellite (top), RGB intensity map (bottom left) and RBF-NN approximation with 300 neurons (bottom right)*

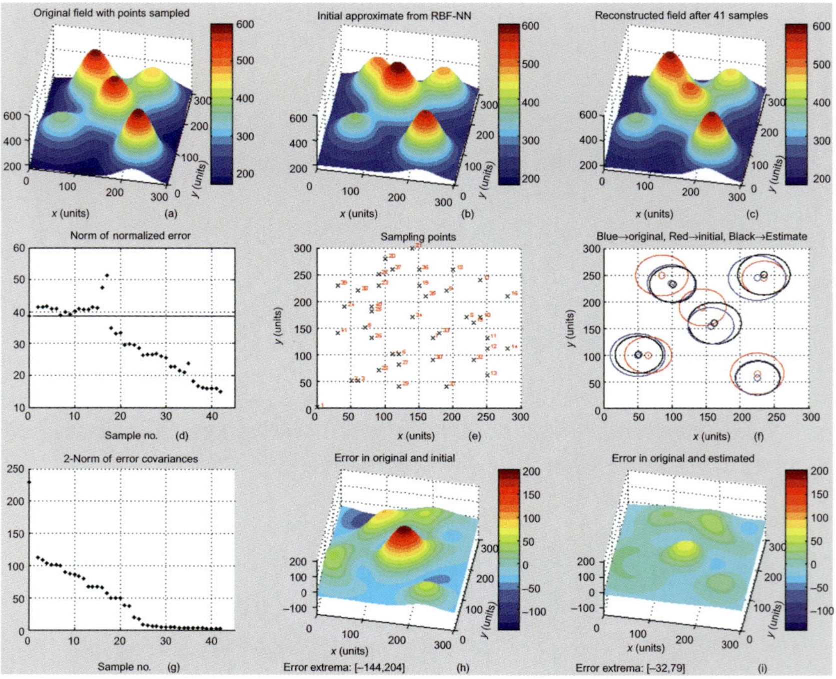


*Figure 4.17 (a–i) Simulation results with GAS for sum of Gaussians stationary field*

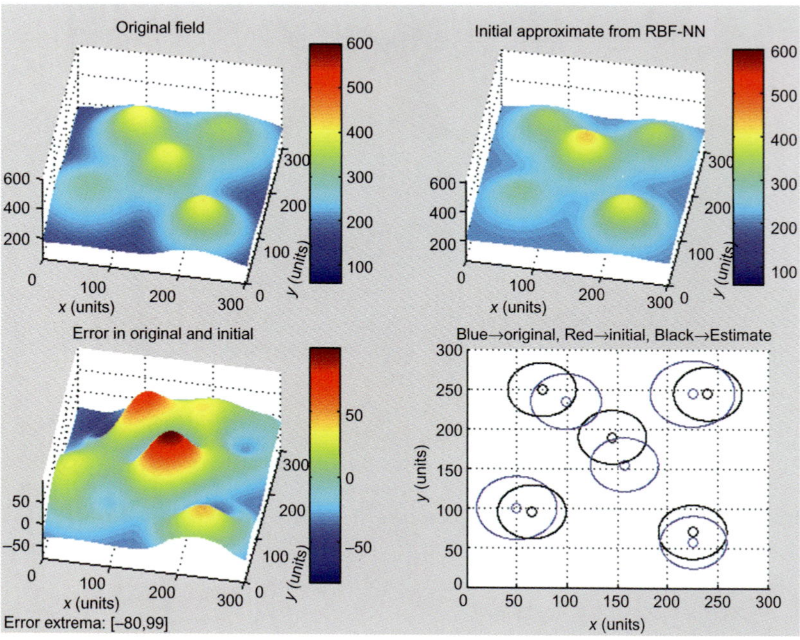


*Figure 4.23 Actual and initial estimated fields from NN for sum of Gaussians time-varying field*

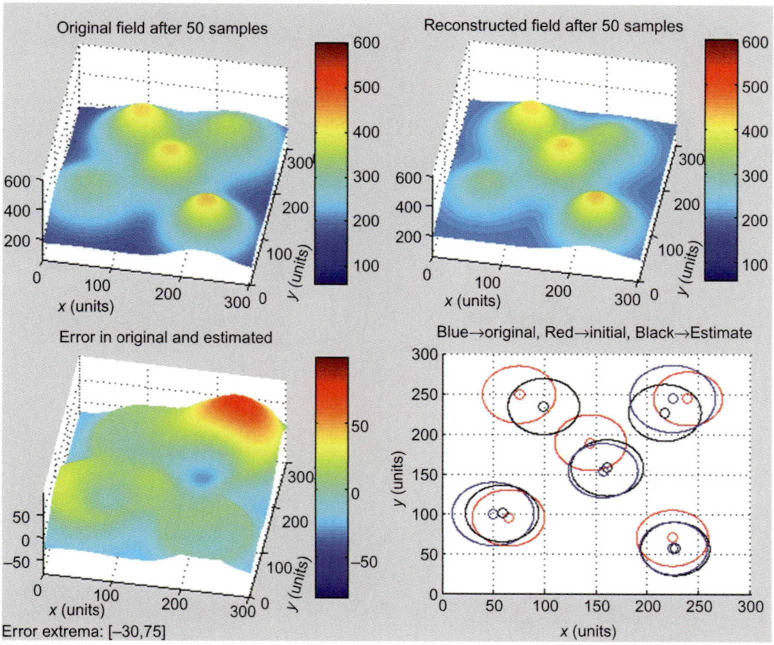


*Figure 4.24 Actual and estimated fields after 50 samples for sum of Gaussians time-varying field*

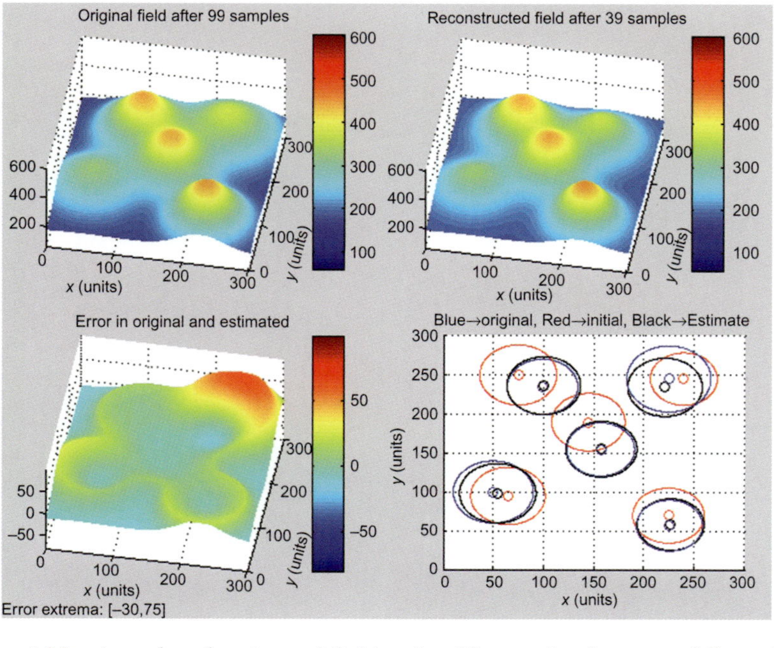


*Figure 4.25 Actual and estimated fields after 99 samples for sum of Gaussians time-varying field*

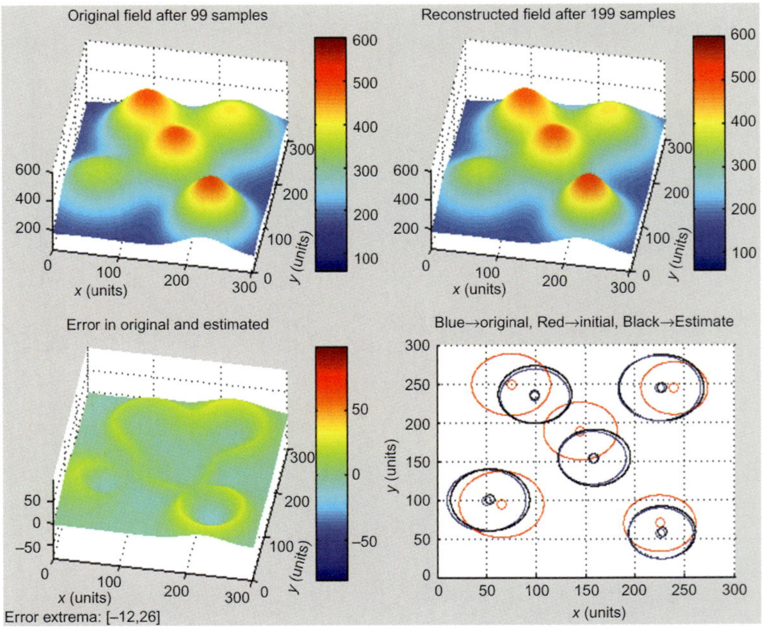


*Figure 4.26 Actual and estimated fields after 199 samples for sum of Gaussians time-varying field*

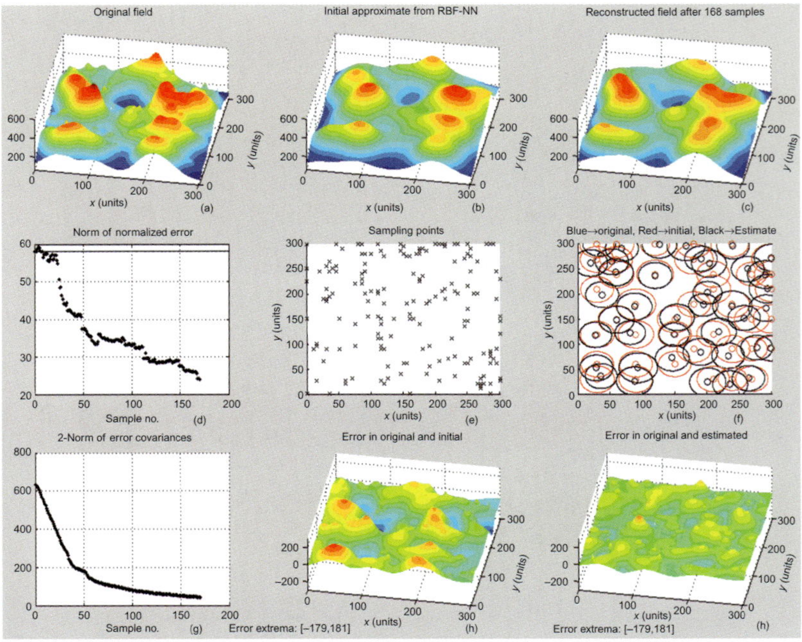


*Figure 4.28 (a–i) Simulation results for GAS for complex time-varying field*

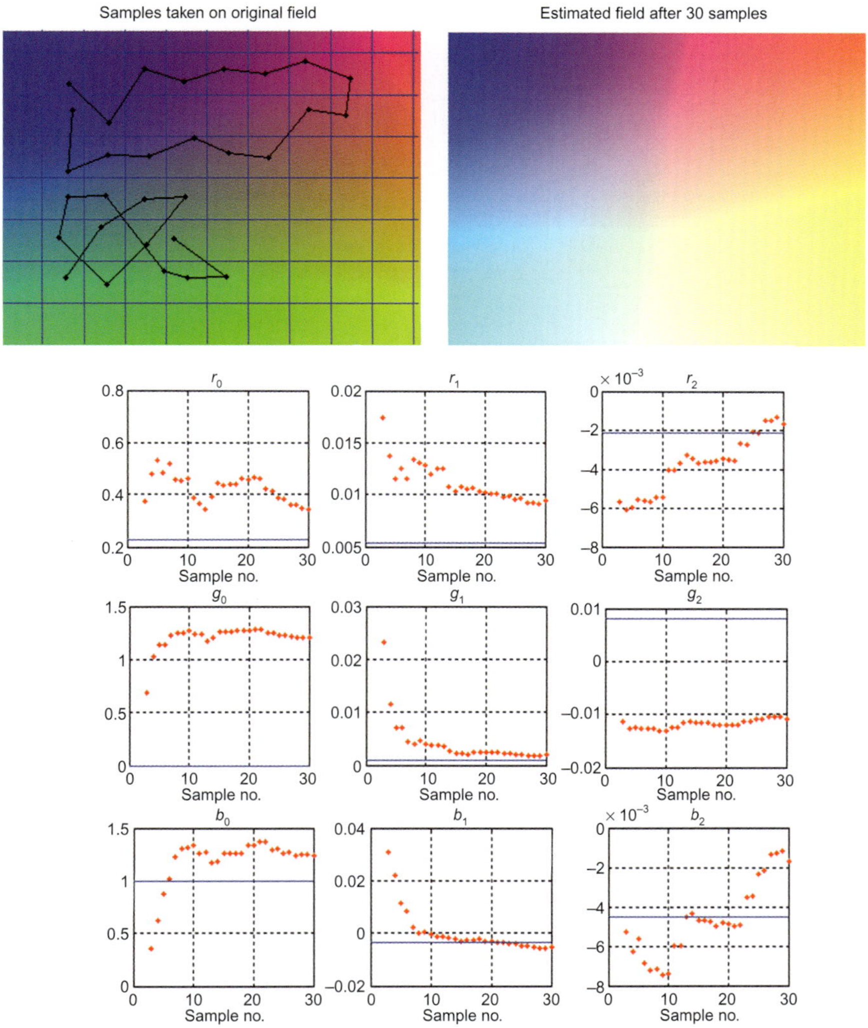


*Figure 5.10 Experimental results for GAS with two robots showing the estimated field after 30 combined samples*

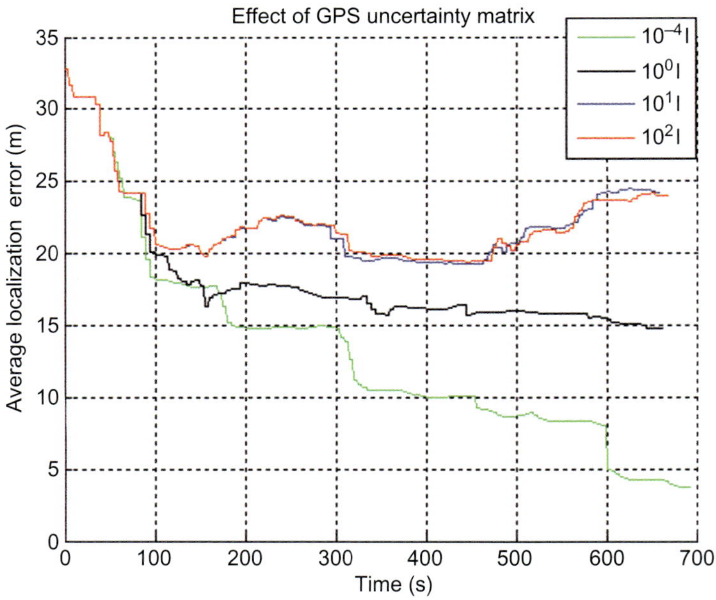


*Figure 7.19 Effect of GPS uncertainty matrix on sensor network localization*

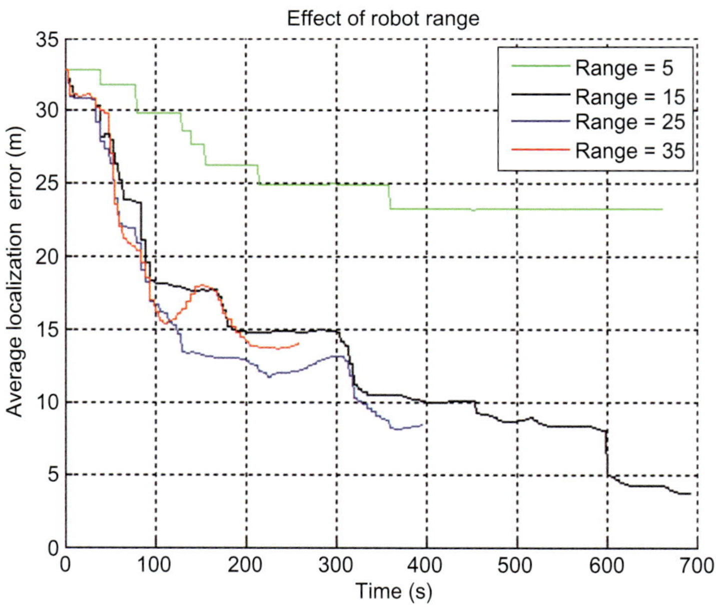


*Figure 7.21 Effect of robot communication range on sensor network localization*

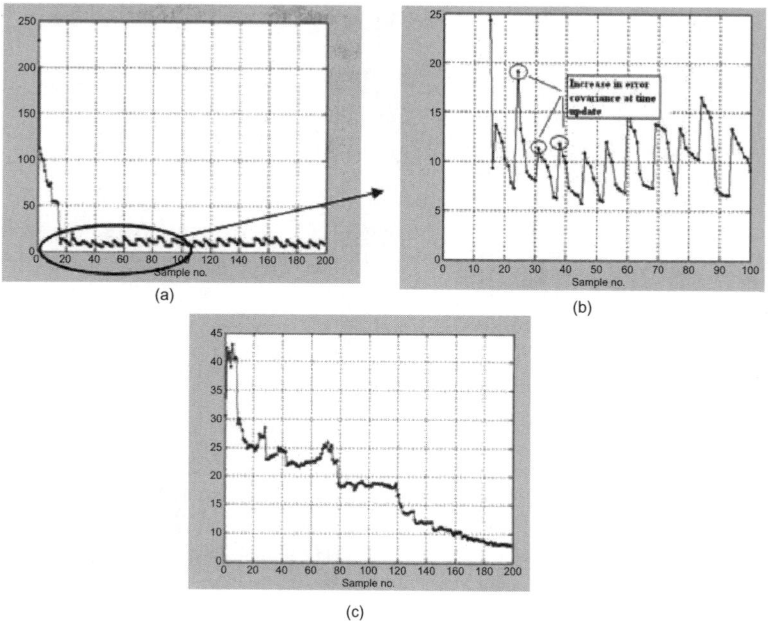


*Figure 4.27* *(a) 2-Norm of error covariance, (b) increase in error covariance when time update occurs and reduces again when measurements are taken and (c) 2-norm of error in actual and estimated fields*

vary as the complexity of the field varies, and the goal of sampling is to minimize the SSE. The numerical values for uncertainties in our model were

$$\begin{array}{lllll} P_{b0} = 100, & P_{a0} = 5, & P_{\sigma 0} = 10^{-7}, & P_{x00} = 1, & P_{y00} = 1 \\ Q_{b0} = 0.02, & Q_{a0} = 0.02, & Q_{\sigma 0} = 0, & Q_{x00} = 0, & Q_{y00} = 0 \\ R = 1 & & & & \end{array} \quad (4.26)$$

Figure 4.28 (see p. C14) shows simulation results for GAS illustrating (a) actual field generated using CA, (b) initial approximate with 40-neuron RBF-NN and spread factor of 30 when grid size of $n = 30$ is used for low-resolution sampling, (c) reconstructed field after 168 samples with GAS heuristic sampling approach when grid size of $p = 5$ is used for high-resolution sampling, (d) SSE in actual and estimated fields that drops faster compared with raster scanning, (e) sampled points, (f) initial and estimated Gaussian locations indicated by solid line and black circles, respectively, (g) 2-norm of error covariance of parameter estimates, which drops faster compared with raster scanning, (h) error in actual and initial estimates and (i) error in actual and final estimates after 168 samples.

Figure 4.29 shows simulation results for raster scan sampling illustrating (a) actual field generated using CA, (b) initial approximate with 40-neuron RBF-NN and spread factor of 30 when grid size of $n = 30$ is used for low-resolution sampling, (c) reconstructed field after 951 samples with raster scanning when grid size of

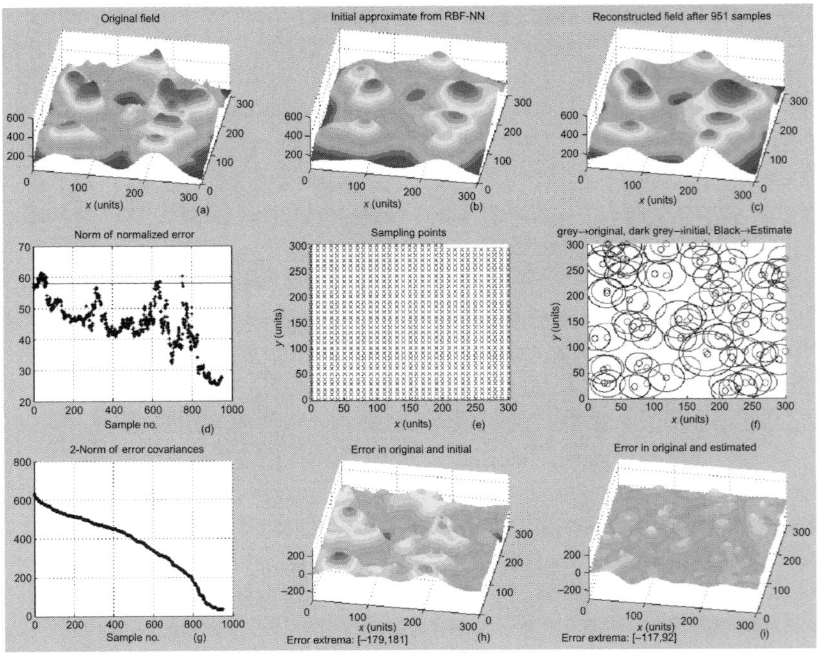


*Figure 4.29 (a–i) Simulation results for raster scan sampling for complex time-varying field*

$p = 10$ is used for high-resolution sampling, (d) SSE in actual and estimated fields, (e) sampled points, (f) initial and estimated Gaussian locations indicated by solid line and black circles, respectively, (g) 2-norm of error covariance of parameter estimates, (h) error in actual and initial estimates and (i) error in actual and final estimates after 951 samples.

Figure 4.29 shows the raster scanning simulation results when sampling is performed row-by-row in a grid size of $p \times p = 20 \times 20$. Figure 4.28 shows the GAS sampling simulation results when a grid size of $p = 5$ and a horizon size of 20 are considered. Grid size is an important parameter. A very large grid size will not reduce the error significantly, even if the whole area is scanned, because thorough

*Table 4.3 Comparison of raster scan and greedy AS for complex time-varying field*

| | **Greedy AS** | **Raster scan** |
|---|---|---|
| Distance | 3380 | 19,020 |
| No. of samples | 168 | 951 |
| Initial 2-norm of error in actual and estimated fields | 58.07 | 58.07 |
| Final 2-norm of error in actual and estimated fields | 23.95 | 27.49 |
| Initial 2-norm of error covariance | 632.5 | 632.5 |
| Final 2-norm of error covariance | 43.23 | 33.62 |

sampling is required in high-variance areas. As the robot starts sampling with given initial estimates and uncertainties, the uncertainty of the parameters does not decrease until the robot reaches the area where those parameters have a significant influence. In other words, the uncertainty of the Gaussian is most reduced when sampling is performed within a few variance values away from its centre.

A comparison of GAS and raster scanning is summarized in Table 4.3. We notice that since raster scan performs a row-by-row scanning, it takes a longer time and many more samples than GAS. The simulation stops when 2-norm of error in actual and estimated fields reduces below a threshold. It can be seen from Figures 4.28 and 4.29 that SSE between the actual and the estimated fields, as well as the 2-norm of state error covariances, decreases faster in case of GAS.

### *4.5.6 Potential fields for safe trajectory generation*

Simulations were also performed to generate fire-safe paths through the estimated field, as shown in Figure 4.30, as presented in section 4.4. In our simulations, we assumed that four fire ellipses are ignited simultaneously, and that four robots are simultaneously sampling in designated areas. Every robot runs a separate EKF-based AS algorithm to estimate the parameters of its local fire field. Fire field data is then aggregated in a central processing location, which is also responsible for online fire-safe trajectory generation. The trajectory is dynamically updated from $X_i$ to $X_{goal}$ every time the field parameter estimate updates. In a practical

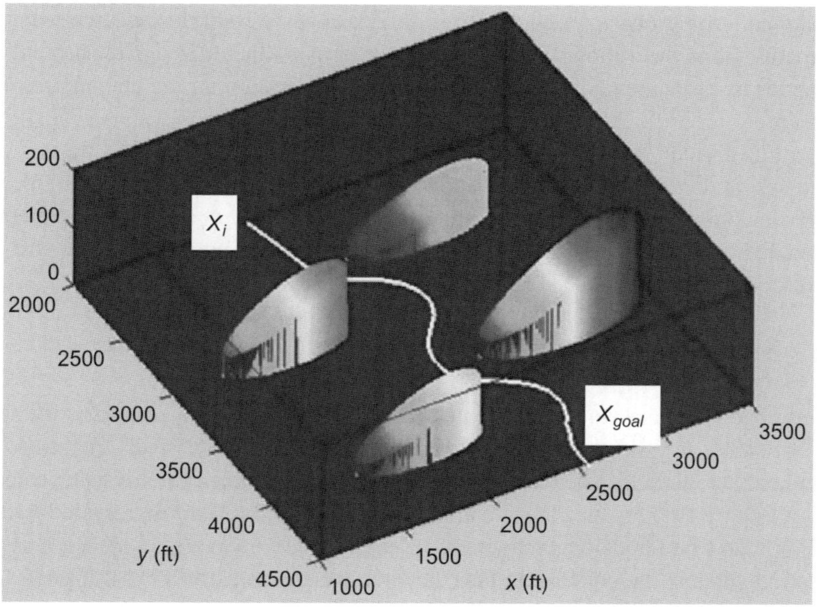


*Figure 4.30 Estimated fire ellipses and dynamically generated path using potential fields for the human firefighter to go from $X_i$ to rescue location $X_{goal}$*

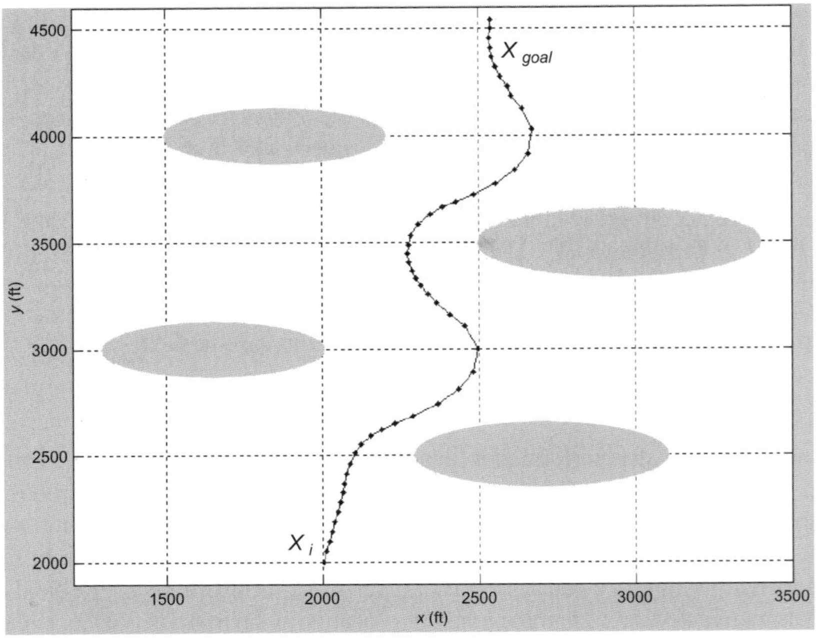


*Figure 4.30 Continued*

implementation scenario, a human firefighter can carry a wireless device receiving estimates from the robots in order to generate a collision-free path around fire obstacles towards the goals using (4.19) and (4.20). Assuming the human crew is at location $X_i = (2000, 2000)$ and needs to go to rescue location $X_{goal} = (2500, 4500)$, the path is divided into 50 segments, and repulsive forces from each of these virtual obstacles and attractive force towards the goal are calculated. The numerical coefficients used in the simulation are $\xi = 1$, $\lambda = 10^6$ and $\delta = 0.01$. The trajectory is updated using this suitable $\delta$ and net force $\vec{F}_p$ value using (4.21), and a fire-safe trajectory is generated.

## 4.6 Summary

In this chapter, a case study is presented for mapping forest files and extensive simulation results to validate the proposed algorithm. We progressively increased the complexity of the field distribution from simple spatial stationary fields represented by only a few parameters to a fairly complex spatio-temporal field represented with an RBF-NN and hundreds of parameters. In each case, convincing evidence was presented to support the conclusion that the EKF-based AS algorithms using the GAS heuristic and an RBF-NN approximation perform more efficiently than a simple raster scan, and that the robot uncertainty can be reduced by sampling.

*Part III*

# Multi-resource strategies

# Chapter 5
# Distributed processing for multi-robot sampling

If the environment is large, it may be impractical for a single robot to navigate to many sampling locations, even in the presence of efficient sampling algorithms. Using multiple robots, the sampling area can be divided into smaller regions, thus reducing the navigation time. Furthermore, since complex fields are represented by hundreds of parameters, it is computationally cumbersome for a single robot to compute and store all parameter estimates and the uncertainty measures. It also quickly becomes unfeasible for individual robots to run a large extended Kalman filter (EKF)-adaptive sampling (AS) algorithm, and share large covariance matrices wirelessly. Furthermore, with multi-robot sampling, the resources can be allocated efficiently if some resources are busy or not available.

If we somehow can distribute the filter computation among multiple robots, the number of computations performed by all the robots will be greater than the processing by just one robot doing sampling. However, we expect that the speed of convergence and reduction in complexity that will be gained is significant. With a single robot, the total field estimation time includes the time necessary for navigation, sensing and computations of the estimate (as there is no communication). With multiple robots, the field estimation time includes the time taken for sensing, computation, communication and final fusion to recover the field. We expect that the speed of convergence increases using multiple robots simply because of sampling in parallel, and the navigation time reduces significantly compared with modest increases in computation, communication and fusion.

Our proposed parametric AS algorithm, EKF-NN-GAS, represents a complex field with sum of overlapping Gaussians, which means that each sampling instance in a region gains information about the parameters that have dominant effect in that region. Therefore, in order to distribute computations, we need to fuse the parameter estimates and construct the map of density distribution.

This problem is similar to reformulating the algorithm from a conventional single-sensor, single-processor system to a multi-sensor, multi-processor system. Distributed algorithms have been used before in many applications, and the degree of parallelism varies from algorithm to algorithm. Examples include target location estimation using several sensors and fusing the measurement either at the central station or at each sensor depending on the degree of parallelism of the multi-sensor fusion algorithm [102]. Another example is formation control for multiple robots [103].

As discussed in Chapter 1, decentralized Kalman Filter and distributed Kalman filter (DKF) are two different problems. In a decentralized algorithm, the filter is full-order, which means that every local filter carries partial information about all parameters, and the information is shared to reach consensus. The objective of distributed algorithms is to efficiently decompose the full-order filter into several reduced-order filters, in order to reduce the computational complexity and communication overhead, and hence improve the scalability. It can be said that decentralization is the first step towards efficient distribution. Decentralized approaches are good enough for applications involving a small number of states such as tracking of objects, etc. But problems such as parametric sampling involve hundreds of parameters, and hence distributing the filter is very important.

This chapter focuses on examining completely and partially centralized, decentralized and distributed multi-robot computations, and formulating a sampling scheme that is efficient for running the proposed multi-robot AS algorithm. Sections 5.1–5.3 summarize the existing approach for completely centralized, completely decentralized and partially centralized federated filters, respectively. Section 5.4 presents the distributed federated Kalman filter (DFKF) along with description of partitioning of sample space, and the distribution of computations and reduction of communication overhead. Finally, in sections 5.5 and 5.6, we discuss the simulation and experimental results, respectively, to demonstrate the effectiveness of the distributed AS algorithm.

## 5.1 Completely centralized filter

In a completely centralized sampling approach, each robot $j$ takes sensor measurement $Z_{j,k+1}$ and transmits it to the central processor, which then calculates parameter estimates $\hat{A}_{k+1}$ and error covariances $P_{k+1}$. The central processor computes the estimated error covariance and parameter estimation using an equation similar to (3.14), with the difference that here it fuses multiple measurements from $N$ robots:

$$P_{k+1} = \left[(P_{k+1}^{-})^{-1} + G_{k+1}^{\mathrm{T}} R_{k+1}^{-1} G_{k+1}\right]^{-1} = \left[(P_{k+1}^{-})^{-1} + \sum_{j=1}^{N} G_{j,k+1}^{\mathrm{T}} R_{j,k+1}^{-1} G_{j,k+1}\right]^{-1}$$

$$\hat{A}_{k+1} = \hat{A}_{k+1}^{-} + P_{k+1} G_{k+1}^{\mathrm{T}} R_{k+1}^{-1} \left[\begin{pmatrix} Z_{1,k+1} \\ Z_{2,k+1} \\ \vdots \\ Z_{N,k+1} \end{pmatrix} - \begin{pmatrix} g_{k+1}(\hat{A}_{k+1}^{-}) \\ g_{k+1}(\hat{A}_{k+1}^{-}) \\ \vdots \\ g_{k+1}(\hat{A}_{k+1}^{-}) \end{pmatrix}\right] \tag{5.1}$$

Figure 5.1 illustrates the completely centralized approach, in which all robots transmit their sensor measurement to the central filter, which then calculates the field estimate using (5.1). This type of scheme is simple, as there is little communication involved and no redundant computations. But the disadvantage is that the sensing robots do not carry any information. Hence, they do not know where to

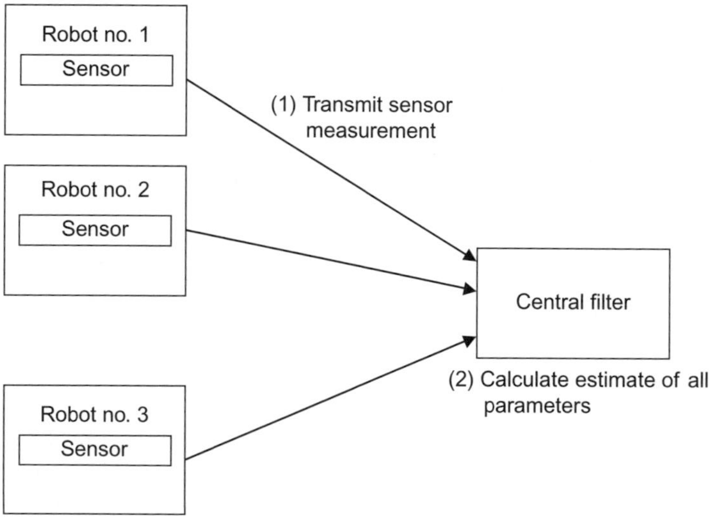


*Figure 5.1 Completely centralized AS algorithm*

sample next until told to do so by the central processor, which is running a large algorithm receiving asynchronous data.

## 5.2 Completely decentralized filter

For a completely decentralized filter implementation, each robot runs the AS algorithm and generates new sampling locations within the vicinity of its current position. The robots take new measurements and calculate only partial estimates of the field parameters and error covariance. After a few samples, the robots communicate and share their field estimate information. The parameter estimate and the error covariance are the two terms each robot needs to transmit to others. Each robot assimilates the received information using, for instance, a decentralized EKF scheme similar to Reference 102.

If a completely decentralized approach is considered, then an AS algorithm running on each robot carries the information about all the field parameters, and there is no data fusion filter. Hence, each robot $i$ can calculate the local field estimate $\hat{A}_{i,k+1,LE}$ and $P_{i,k+1,LE}$ using the following equations:

$$
\begin{aligned}
P_{i,k+1,LE} &= [(P^{-}_{i,k+1})^{-1} + G^{\mathrm{T}}_{i,k+1}R^{-1}_{i,k+1}G_{i,k+1}]^{-1} \\
\hat{A}_{i,k+1,LE} &= \hat{A}^{-}_{i,k+1} + P_{i,k+1,LE}G^{\mathrm{T}}_{i,k+1}(R_{i,k+1})^{-1}[Z_{i,k+1} - g_{k+1}(X_{k+1}, \hat{A}^{-}_{k+1})]
\end{aligned} \tag{5.2}
$$

Robot $i$ acquires the local estimate $\hat{A}_{j,k+1,LE}$ and $P_{j,k+1,LE}$ from the other $(N - 1)$ robots and assimilates it to get the complete information $P_{i,k+1}$ and $\hat{A}_{i,k+1}$. The following assimilation equation needs to run on robot $i$:

$$(P_{i,k+1})^{-1} = (P^{-}_{i,k+1})^{-1} + \sum_{j=1}^{N}[(P_{j,k+1,LE})^{-1} - (P^{-}_{j,k+1})^{-1}]$$

$$\hat{A}_{i,k+1} = P_{i,k+1}\left[(P^{-}_{i,k+1})^{-1}\hat{A}^{-}_{i,k+1} + \sum_{j=1}^{N}[(P_{j,k+1,LE})^{-1}\hat{A}_{j,k+1,LE} - (P^{-}_{j,k+1})^{-1}\hat{A}^{-}_{j,k+1}]\right] \tag{5.3}$$

Figure 5.2 illustrates the completely decentralized filter in which each robot has a local filter to compute local estimates and a global filter to assimilate the estimates acquired from other nodes and generate the global field estimate.

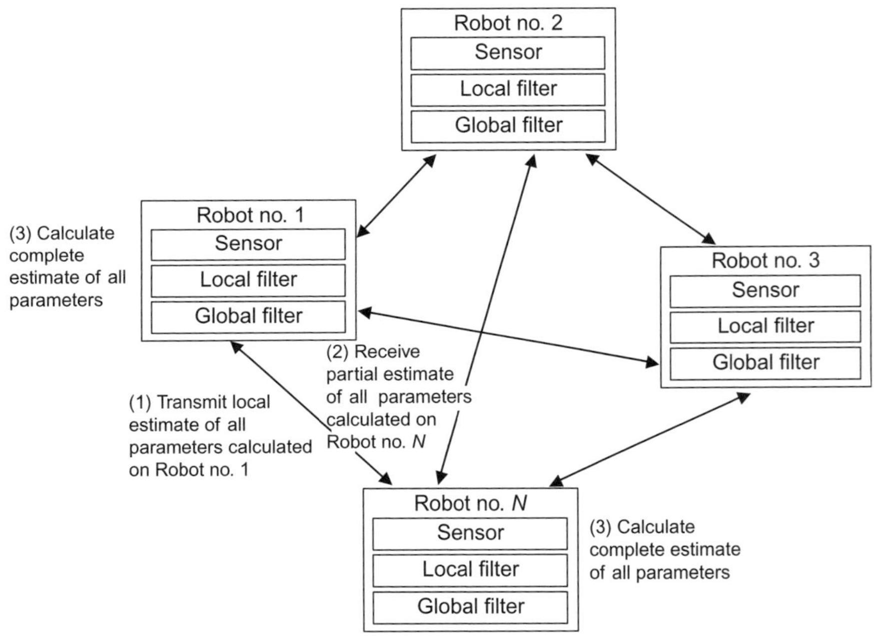


*Figure 5.2 Completely decentralized AS algorithm*

The advantage of this approach is that it does not involve any approximations, and there is no dependence on a central filter, at least for computing the local estimate. Also the objective of sampling in parallel can be successfully achieved. The disadvantage of the algorithm is that it is inefficient in terms of communication and computational requirements. The network has to be fully connected, and there is excessive communication. This type of algorithm is good for applications such as target tracking because the problem involves estimating only a few parameters (such as location, speed, etc., of the target). Hence, the distribution of computations may not be that important if the number of parameters is small. For an AS algorithm running such that each robot has a designated sampling region, different robots carry better information about different parameters. Hence, a loss of information from only a few robots will make a significant difference to the field estimation result. This is not the case for target tracking, where a loss of information from one or two nodes does not significantly impact the accuracy of estimate, if enough nodes can still track the target.

For an AS scenario, if, for example, a field is represented by 100 parameters, running this decentralized algorithm would require each robot to calculate the local estimate of 100 parameters, and to wirelessly transmit an error covariance matrix of size $100 \times 100$ and a parameter estimate vector of size $100 \times 1$ to every other robot. Clearly, such a scheme would be very inefficient and not scalable.

## 5.3 Partially centralized federated filter

A trade-off approach can be adopted for sampling, in which each robot takes sensor measurements, estimates local error covariances and field parameters, and transmits this information to the global filter for assimilation, in a similar fashion to the approach proposed in Reference 104. Each robot runs (5.2), but the fusion is done only at the fusion filter using (5.3).

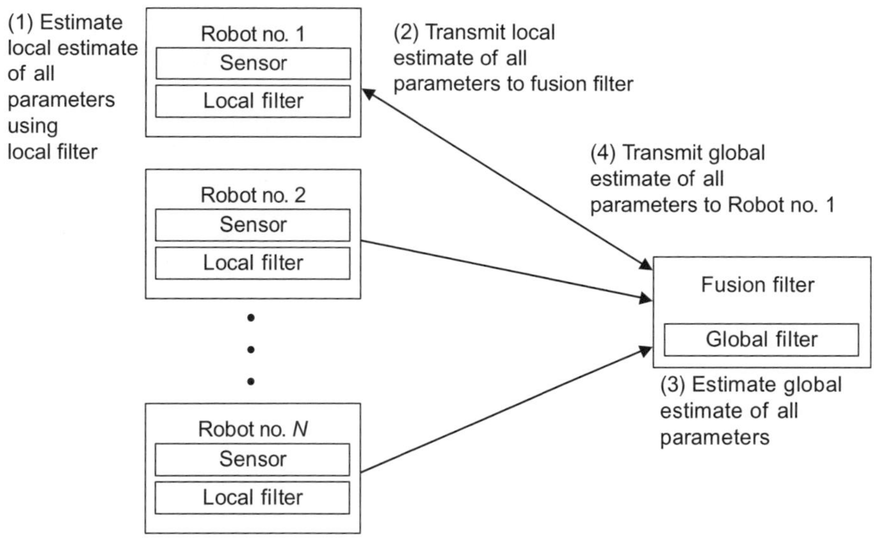


*Figure 5.3 Partially centralized federated AS algorithm*

Figure 5.3 illustrates the centralized federated filter in which each robot calculates a local estimate and transmits it to the fusion filter, which then computes the global field estimate. The advantage of this approach is that there is less communication compared with the completely decentralized case. Although none of the robots carry complete information about all parameters, this approach is more efficient than the centralized implementation because AS algorithms can run based on the knowledge about local parameter estimates. The formulation of the filter can be implemented in a general approach by assuming that each robot carries some information about all the parameters. For multi-robot sampling performed in such a way that sampling area for robots is assigned *a priori*, it is quite reasonable to assume that every robot carries information about some parameters and no information about others. This assumption can be used to distribute the filter equations, as described in the following section.

## 5.4 Distributed federated Kalman filter

In this section, we describe an algorithm to distribute the computations of the field parameter estimates among multiple robots, and we show that the scheme is efficient in terms of communication and computation overhead. The sampling area is divided such that each robot is responsible for sampling certain regions and sets of parameters. A block diagram of the distribution scheme is shown in Figure 5.4. Compared to the approach presented in previous section, such a distributed scheme involves the transmission and computation of only certain parameters.

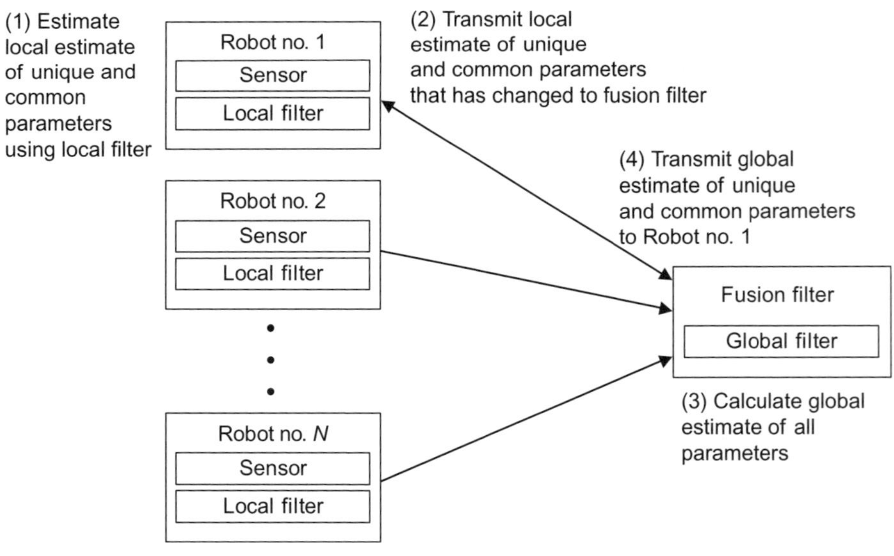


*Figure 5.4 Distributed federated AS algorithm*

### 5.4.1 Partitioning of sampling area

A method is clearly needed to efficiently divide the sampling area into clusters, in order to run a parallel AS algorithm with multiple robots. Fuzzy *c*-means clustering (FCM) has been used in many applications for classification of numerical data [74]. The basics of FCM were presented in Chapter 1. Moreover, centroidal Voronoi tessellation (CVT) diagrams [105] have been recently proposed for forming non-uniform-sized grids to better explore high-variance areas for non-parametric distributions [11]. Here a scheme is proposed to efficiently divide the sampling area for parametric distributions using both FCM and CVT. In this approach, the FCM algorithm clusters based on estimated location of the centre of approximating Gaussians.

The implementation of the CVT diagram is based on Lloyd's algorithm [105] and uses the centroid locations acquired by fuzzy clustering to fuse all points in discrete space that are closest to the centroid as a single group. Mathematically, a point $p$ on the field is part of the cluster $r$ if

$$|p - c_r| \leq |p - c_s|, \quad s = 1, \ldots, C, \quad s \neq r \tag{5.4}$$

As a result, more Gaussians will overlap in those areas where there are large field variations. The use of FCM and the CVT diagram for area classification results in regions that have more variations to be as small as required in order to sample them thoroughly. The area with less variation, though large, requires fewer samples, since it is represented by only a few parameters. The idea is illustrated in the simulation of Figure 5.5, where we partition the sampling space of a non-uniform distribution represented by $L = 100$ Gaussians into eight regions.

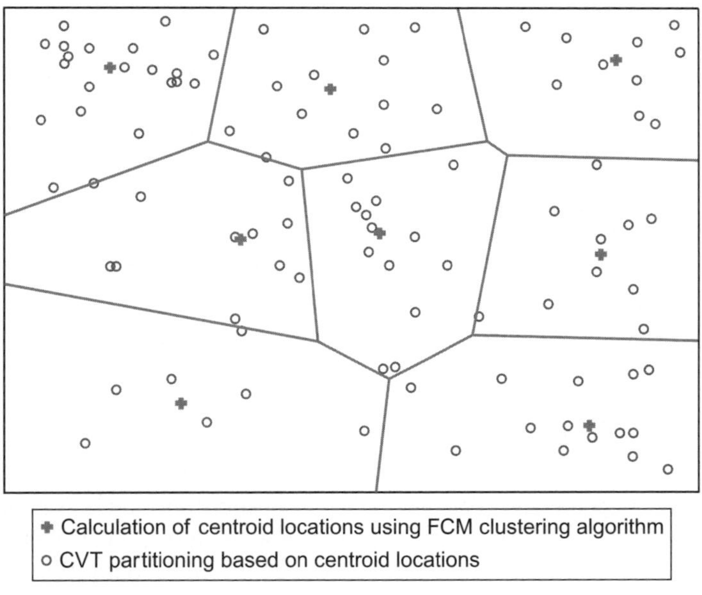


*Figure 5.5 Sampling area with Gaussian field centres partitioning performed in two steps using FCM and CVT*

Individual robots are only interested in the parameters that are either unique to them or common with other robots. These parameters fall under three categories of obvious significance, referred here as *common, unique* and *uncorrelated*. Common parameters are most crucial to the field estimate, as they exist near the border of the sampling areas. The sampling information from multiple robots contributes to the

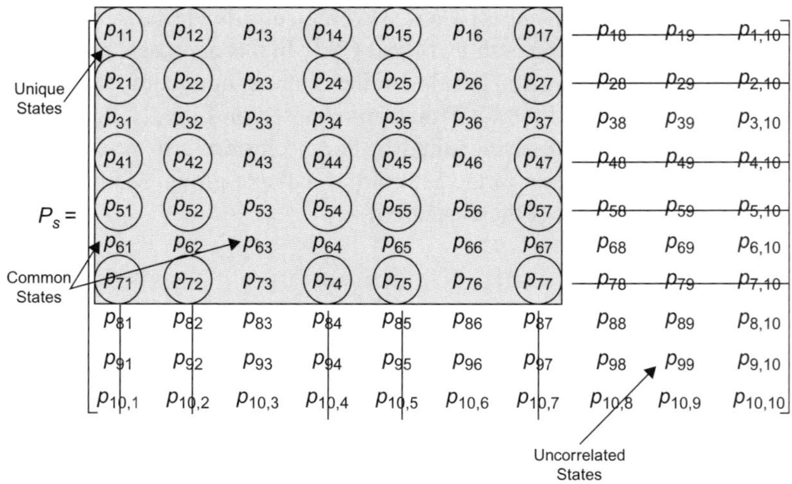


*Figure 5.6 Three categories of parameters (unique, common and uncorrelated) can be formed if DFKF approach is used along with partitioning of sampling area for each robot*

estimation of common parameters. As shown in Figure 5.4, only the estimate of common and unique parameters that have changed needs to be transmitted to the fusion filter. Furthermore, the cross-covariance terms of unique states with uncorrelated states can be assumed to be zero. Similarly, some of the *common* states are not correlated with the *uncorrelated* states.

The idea is further illustrated in Figure 5.6 by taking an example of a region that has two clusters. Cluster 1 has *unique parameters* 1, 2, 4, 5 and 7, and cluster 2 has unique parameters 8, 9 and 10. The *common parameters* are 3 and 6. Therefore, for cluster 1, the uncorrelated parameters are whatever is other than the unique and common parameters.

## 5.4.2 Distributed computations and communications

The objective of the work presented in this section is to modify the formulation of a completely decentralized federated scheme, in order to reduce communication and computational load. This formulation is new because it considers the cross-covariance terms of neighbouring Gaussians and ignores the ones that are far from each other as a trade-off between accuracy and computational complexity. Accurate DKF is not possible in this AS problem because local measurement models are not available. Furthermore, the use of global measurement models on each node requires the estimate of all parameters, which will contradict the motivation behind the implementation of DKF. There are other schemes that handle the error covariance terms 'very lightly' such as Kalman consensus schemes, which take the average of error covariance of parameter estimate in order to implement the DKF with only neighbouring node communication [106].

*Table 5.1 Comparison of computational complexity and communication overhead for centralized, decentralized, federated and distributed federated filter (DFKF)*

| | Computations | | | Communication |
|---|---|---|---|---|
| | **Robot** | **Fusion centre** | **Combined** | |
| Centralized filter | – | $O(Nmn^3)$ | $O(Nmn^3)$ | $O(Nm)$ |
| Decentralized filter | $O(mn^3 + (N-1)pn^3)$ | – | $O(Nmn^3 + N(N-1)pn^3)$ | $O(N(N-1)pn^3)$ |
| Decentralized federated filter | $O\left(\frac{mn^3}{N^3}\right)$ | $O(pn^3)$ | $O\left(\frac{mn^3}{N^3} + pn^3\right)$ | $O\left(\frac{pn^3}{N^2}\right)$ |

EKF has $O(n^3)$ computational complexity if each sample updates all the $S$ parameters of the $n$-dimensional parametric field. However, it can be assumed that a single sample affects only neighbouring parameters. With this assumption, the algorithm can run in distributed fashion, and sampling node computation complexity can be reduced. Only the fusion filter's complexity remains of order $O(n^3)$ because it needs to combine information about all the parameters. However, this central field parameter fusion process occurs less frequently.

Table 5.1 illustrates a comparison of computations and communication complexity for centralized, completely decentralized, federated decentralized and distributed filter. Let $N$ be the number of sampling robots, $m$ be the number of sensor measurements per robot, $n$ be the number of field parameters and $p$ be the number of times robots communicate to share their information.

For the centralized filter, the sensing robots do not perform any computation. Hence, the computational and communication complexity are $O(Nmn^3)$ and $O(Nm)$, respectively.

For a completely decentralized filter, the computational complexity to calculate the local estimate on each robot is $O(mn^3)$, and to calculate the global estimate on each robot is $O((N-1)pn^3)$, after taking estimates from $(N-1)$ robots at a frequency of $p$. Hence, the combined computational complexity becomes $O(Nmn^3 + N(N-1)pn^3)$. At the same time, the communication complexity is $O(N(N-1)pn^3)$.

In order to reduce the communication overhead and computational complexity, a federated filter calculates the global estimate on the fusion filter, which reduces the computational complexity to $O(mn^3 + pn^3)$, and the communication complexity to $O(Npn^3)$.

Finally, for the proposed distributed version of federated decentralized filter, instead of calculating $n$ states on a single robot, we simply calculate the estimate of $n/N$ states on a single robot. This approach reduces the computational and communication complexity to $O(mn^3/N^2 + pn^3)$ and $O(pn^3/N^2)$, respectively.

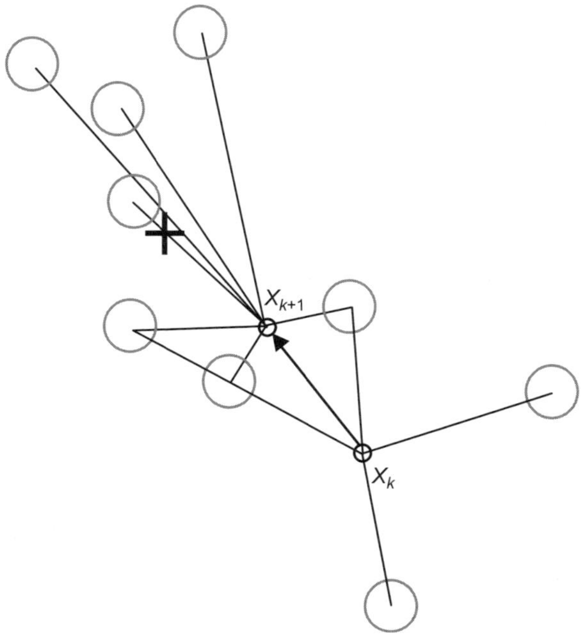


*Figure 5.7 Centres of Gaussians that are close to sampling locations $x_k$ and $x_{k+1}$*

Let us now assume that the field classification includes a large cluster for a continuous field distribution. A *large cluster* is defined as the one whose parameters are independent of parameters outside that cluster. This means that the Gaussians representing this large cluster do not overlap with other Gaussians outside the cluster. Furthermore, let us assume that there are $L$ parameters in a large cluster, and that a cluster head robot is in charge of carrying the updated parameter estimates and the error covariance of all $L$ parameters.

The large cluster will contain smaller clusters. A *small cluster* is defined as one whose parameters are estimated by the same robot for a span of time until the tessellation changes. The participant parameters, and hence the shape of a small cluster, might change after some samples are taken and as the parameter estimates change. Parameters of one small cluster can be dependent on parameters of another small cluster around it. Furthermore, let us assume that there are $S$ parameters in or around a particular small cluster that are expected to change because of sampling for next few samples by a particular robot. As shown in an example in Figure 5.7, sampling at $x_k$ only provides information about the neighbouring five Gaussians. Similarly, $x_{k+1}$ provides information about the neighbouring seven Gaussians, three of which are common between the two sampling locations.

Assume $M$ parameters are expected to change by sampling a particular location. In our scheme, instead of computing the error covariance and parameter estimate of size $S$, each sample updates only $M$ parameters.

Summarizing the clustering parameters, and referring to the diagram in Figure 5.8, let

$N$: number of sampling robots
$L$: parameters in large cluster
$S$: parameters in or around small cluster
$M$: parameters that are expected to change by sampling a particular location
$C$: parameters whose estimates have changed since last update from cluster head
$P_L$: error covariance of $L$ parameters that includes unique, common and uncorrelated parameters
$P_S$: error covariance of $S$ parameters that are either unique or common
$P_M$: error covariance of $M$ parameters
$P_C$: error covariance of $C$ parameters
$\hat{A}_L$: estimate of $L$ parameters
$\hat{A}_S$: estimate of $S$ parameters
$\hat{A}_M$: estimate of $M$ parameters
$\hat{A}_C$: estimate of $C$ parameters
$P_L$ and $\hat{A}_L$ are carried by the cluster head, $M \subset C \subset S \subset L$
$g_L(X_0, A_{L,0})$: sum of $L$ Gaussians
$g_S(X_0, A_{S,0})$: sum of $S$ Gaussians
$g_M(X_0, A_{M,0})$: sum of $M$ Gaussians

Sampling in region $M$, the field measurements are given by

$$\begin{aligned} Z_{k+1} &= g(X_{k+1}, A_{L,k+1}) + v_{i,k+1} \approx g(X_{k+1}, A_{S,k+1}) + v_{i,k+1} \\ &\approx g(X_{k+1}, A_{M,k+1}) + v_{i,k+1} \end{aligned} \tag{5.5}$$

$$g(X_{k+1}, \hat{A}_{M,k+1}) = \sum_{u=1}^{M} g(X_{k+1}, \hat{A}_{u,k+1}) \tag{5.6}$$

where $M$ Gaussians are the ones whose centres are closest to the currently sampled location $X_{i,k+1}$ by robot $i$, and

$$G_{M,k+1} = \frac{\partial g(X_{k+1}, A_{M,k+1})}{\partial \hat{A}_{M,k+1}} \in \Re^{1xM}, \quad G_{S,k+1} = \frac{\partial g(X_{k+1}, A_{S,k+1})}{\partial \hat{A}_{S,k+1}} \in \Re^{1xS} \tag{5.7}$$

$g(X_{k+1}, A_{L,k+1})$ is sum of $L$ Gaussians, but for a particular sampling location it will only affect $M$ Gaussians considerably, and to a lesser extent other Gaussians. These $M$ Gaussians can be assumed to be the ones close to the currently sampling location. The dimension of the matrices is reduced by binary transformation matrices $U$ that act as masks over the actual dense EKF matrices. The communication between robots is reduced by transmitting only the parameter values that changed since last update, instead of transmitting the information about all the parameters every time. The transformation matrix is also sent along with the estimates to the fusion filter in order for the cluster head to know which parameters have changed since the last update. The proposed algorithm consists of the following steps:

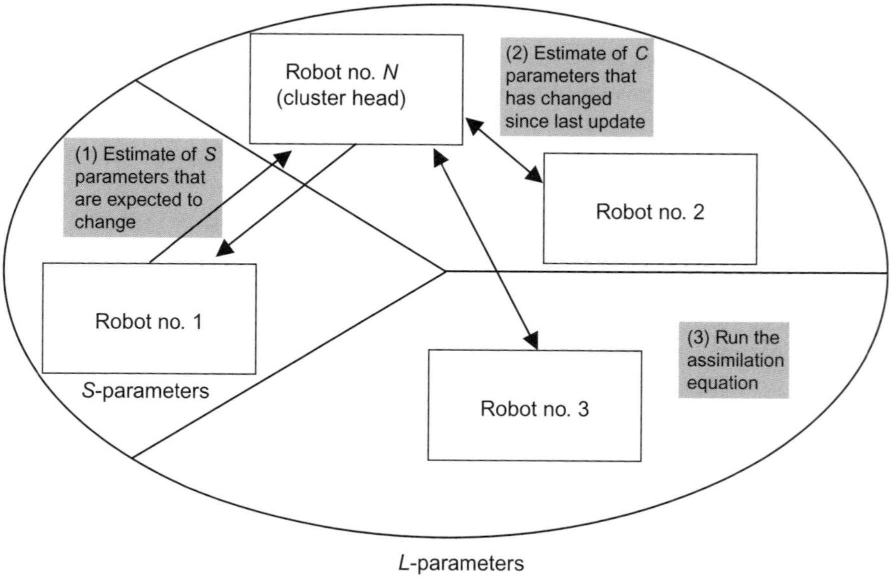


*Figure 5.8 Partially centralized AS algorithm with distributed computations*

*Distributed Federated Kalman Filter*

Step 1: At the cluster head transformation from $(P_{L,k}, \hat{A}_{L,k})$ to $(P_{S,k}, \hat{A}_{S,k})$. The cluster head evaluates the initial estimate of $(P_{S,k}, \hat{A}_{S,k})$ by first generating the transformation matrix $U_{LS,k}$, and transmitting $(P_{S,k}, \hat{A}_{S,k})$ to robot 1. The matrix $U_{LS,k} = U^{\mathrm{T}}_{SL,k}$ is kept in memory by the cluster head for the final assimilation stage:

$$
\begin{aligned}
&P_{S,k} = U^{\mathrm{T}}_{LS,k} P_{L,k} U_{LS,k} \\
&\hat{A}_{S,k} = U^{\mathrm{T}}_{LS,k} \hat{A}_{L,k} \\
&\hat{A}_{S,k} \in \Re^{Sx1}, \quad P_{S,k} \in \Re^{SxS}, \quad \hat{A}_{L,k} \in \Re^{Lx1}, \quad P_{L,k} \in \Re^{LxL}, \quad U_{LS,k} \in \Re^{LxS}
\end{aligned} \tag{5.8}
$$

Step 2: Transmit estimate of *S* parameters $(P_{S,k}, \hat{A}_{S,k})$ to robot 1.

Step 3: Take measurement sample and calculate estimate of *M* parameters $(P_{M,k+1}, \hat{A}_{M,k+1})$ via

$$
P_{M,k+1} = [(U^{\mathrm{T}}_{SM,k+1} P_{S,k} U_{SM,k+1})^{-1} + G^{\mathrm{T}}_{M,k+1} R^{-1}_{k+1} G_{M,k+1}]^{-1} \in \Re^{MxM} \tag{5.9}
$$

$$\hat{A}_{M,k+1} = U_{SM,k+1}^{\mathrm{T}}\hat{A}_{S,k} + P_{M,k+1}G_{M,k}^{\mathrm{T}}(R_{k+1})^{-1}[Z_{k+1} - g(X_k, \hat{A}_{M,k})] \in \Re^{Mx1} \tag{5.10}$$

$$G_{M,k+1} = \frac{\partial g(X_k, \hat{A}_{M,k+1}^{-})}{\partial \hat{A}_{M,k+1}^{-}} \in \Re^{1xM} \tag{5.11}$$

$$\begin{aligned} &\hat{A}_{M,k+1} \in \Re^{Mx1}, \quad P_{M,k+1} \in \Re^{MxM}, \quad \hat{A}_{S,k+1} \in \Re^{Sx1}, \\ &P_{S,k+1} \in \Re^{SxS}, \quad U_{SM,k+1} \in \Re^{SxM} \end{aligned} \tag{5.12}$$

Step 4: Transformation from $(P_{M,k+1}, \hat{A}_{M,k+1})$ to $(P_{C,k+1}, \hat{A}_{C,k+1})$ using

$$\begin{aligned} P_{C,k+1} &= U_{C01}^{\mathrm{T}} P_{C,k} U_{C01} + U_{MC,k+1}^{\mathrm{T}}(P_{M,k+1} - U_{SM,k+1}^{\mathrm{T}} P_{S,k} U_{SM,k+1}) U_{MC,k+1} \\ \hat{A}_{C,k+1} &= U_{C01}^{\mathrm{T}} \hat{A}_{C,k} + U_{MC,k+1}^{\mathrm{T}}(\hat{A}_{M,k+1} - U_{SM,k+1}^{\mathrm{T}} \hat{A}_{S,k}) \end{aligned} \tag{5.13}$$

$$\begin{aligned} &\hat{A}_{C,k+1} \in \Re^{Cx1}, \quad P_{C,k+1} \in \Re^{CxC}, \quad \hat{A}_{C,k} \in \Re^{C_0x1}, \quad P_{C,k} \in \Re^{C_0xC_0} \\ &U_{MC,k+1} \in \Re^{MxC}, \quad U_{C01} \in \Re^{C_0xC} \end{aligned} \tag{5.14}$$

Step 5: Transmit estimate of $(P_{C,k+1}, \hat{A}_{C,k+1}, U_{CS,k+1})$ parameters to cluster head when requested.

Step 6: *At robot 1* – Transformation from $(P_{M,k+1}, \hat{A}_{M,k+1})$ to $(P_{S,k+1}, \hat{A}_{S,k+1})$ for further sampling. Go to Step 3:

$$\begin{aligned} P_{S,k+1} &= P_{S,k} + U_{MS,k+1}^{\mathrm{T}}(P_{M,k+1} - U_{SM,k+1}^{\mathrm{T}} P_{S,k} U_{SM,k+1}) U_{MS,k+1} \\ &= P_{S,k} + U_{MS,k+1}^{\mathrm{T}} G_{M,k+1}^{\mathrm{T}} R_{k+1}^{-1} G_{M,k+1} U_{MS,k+1} \\ \hat{A}_{S,k+1} &= \hat{A}_{S,k} + U_{MS,k+1}^{\mathrm{T}}(\hat{A}_{M,k+1} - U_{SM,k+1}^{\mathrm{T}} \hat{A}_{M,k}) \\ \hat{A}_{S,k+1} &\in \Re^{Sx1}, \quad P_{S,k+1} \in \Re^{SxS}, \quad \hat{A}_{M,k+1} \in \Re^{Mx1}, \quad P_{M,k+1} \in \Re^{MxM}, \\ &\quad U_{MS,k+1} \in \Re^{MxS} \\ U_{MS,k+1} &= U_{SM,k+1}^{\mathrm{T}} \end{aligned} \tag{5.15}$$

Step 7: *At cluster head* – Run the assimilation equations after $n$ steps:

$$\begin{aligned} (P_{L,k+n})^{-1} &= (P_{L,k})^{-1} + \sum_{j=1}^{N} [(P_{L,k+n})^{-1} - (P_{L,k})^{-1}]_j \\ \hat{A}_{L,k+n} &= P_{L,k+n} \left[ (P_{L,k})^{-1} \hat{A}_{L,k} + \sum_{j=1}^{N} [(P_{L,k+n})^{-1} \hat{A}_{L,k+n} - (P_{L,k})^{-1} \hat{A}_{L,k}] \right]_j \end{aligned} \tag{5.16}$$

where

$$[P_{L,k+n}]_j = [P_{L,k} + U^{\mathrm{T}}_{SL,k+n}(U^{\mathrm{T}}_{CS,k+n}P_{C,k+n}U_{CS,k+n} - U^{\mathrm{T}}_{CS,k}P_{C,k}U_{CS,k})U_{SL,k+n}]_j \tag{5.17}$$

$$[\hat{A}_{L,k+n}]_j = [\hat{A}_{L,k} + U^{\mathrm{T}}_{SL,k+n}(U^{\mathrm{T}}_{CS,k+n}\hat{A}_{C,k+n} - U^{\mathrm{T}}_{CS,k}\hat{A}_{C,k})]_j \tag{5.18}$$

*Let us now consider an example for the proposed algorithm:*

Assume that there are ten parameters in a small cluster $S$ to be calculated by a robot. The error covariance $P_{S,k}$ and parameter estimate $A_{S,k}$ matrices can be written as

$$P_{S,k} = \begin{bmatrix} p_{11} & p_{12} & p_{13} & p_{14} & p_{15} & p_{16} & p_{17} & p_{18} & p_{19} & p_{1,10} \\ p_{21} & p_{22} & p_{23} & p_{24} & p_{25} & p_{26} & p_{27} & p_{28} & p_{29} & p_{2,10} \\ p_{31} & p_{32} & p_{33} & p_{34} & p_{35} & p_{36} & p_{37} & p_{38} & p_{39} & p_{3,10} \\ p_{41} & p_{42} & p_{43} & p_{44} & p_{45} & p_{46} & p_{47} & p_{48} & p_{49} & p_{4,10} \\ p_{51} & p_{52} & p_{53} & p_{54} & p_{55} & p_{56} & p_{57} & p_{58} & p_{59} & p_{5,10} \\ p_{61} & p_{62} & p_{63} & p_{64} & p_{65} & p_{66} & p_{67} & p_{68} & p_{69} & p_{6,10} \\ p_{71} & p_{72} & p_{73} & p_{74} & p_{75} & p_{76} & p_{77} & p_{78} & p_{79} & p_{7,10} \\ p_{81} & p_{82} & p_{83} & p_{84} & p_{85} & p_{86} & p_{87} & p_{88} & p_{89} & p_{8,10} \\ p_{91} & p_{92} & p_{93} & p_{94} & p_{95} & p_{96} & p_{97} & p_{98} & p_{99} & p_{9,10} \\ p_{10,1} & p_{10,2} & p_{10,3} & p_{10,4} & p_{10,5} & p_{10,6} & p_{10,7} & p_{10,8} & p_{10,9} & p_{10,10} \end{bmatrix}, A_{S,k} = \begin{bmatrix} a_1 \\ a_2 \\ a_3 \\ a_4 \\ a_5 \\ a_6 \\ a_7 \\ a_8 \\ a_9 \\ a_{10} \end{bmatrix} \tag{5.19}$$

Let the samples taken at times $(k + 1)$, $(k + 2)$ and $(k + 3)$, respectively, estimate the parameters (2, 4, 7), (3, 4, 6) and (1, 2, 4, 9). Then:

$$A_{M,k+1} = \begin{bmatrix} a_2 \\ a_4 \\ a_7 \end{bmatrix}, \quad A_{M,k+2} = \begin{bmatrix} a_3 \\ a_4 \\ a_6 \end{bmatrix}, \quad A_{M,k+3} = \begin{bmatrix} a_1 \\ a_2 \\ a_4 \\ a_9 \end{bmatrix} \tag{5.20}$$

and

$$P_{M,k+1} = \begin{bmatrix} p_{22} & p_{24} & p_{27} \\ p_{42} & p_{44} & p_{47} \\ p_{72} & p_{74} & p_{77} \end{bmatrix}, \quad P_{M,k+2} = \begin{bmatrix} p_{33} & p_{34} & p_{36} \\ p_{43} & p_{44} & p_{46} \\ p_{63} & p_{64} & p_{66} \end{bmatrix}$$
$$P_{M,k+3} = \begin{bmatrix} p_{11} & p_{12} & p_{14} & p_{19} \\ p_{21} & p_{22} & p_{24} & p_{29} \\ p_{41} & p_{42} & p_{44} & p_{49} \\ p_{91} & p_{92} & p_{94} & p_{99} \end{bmatrix} \tag{5.21}$$

The transformation matrices at these instances will be

$$U_{SM,k+1} = \begin{bmatrix} 0&0&0\\1&0&0\\0&0&0\\0&1&0\\0&0&0\\0&0&0\\0&0&1\\0&0&0\\0&0&0\\0&0&0 \end{bmatrix}, \quad U_{SM,k+2} = \begin{bmatrix} 0&0&0\\0&0&0\\1&0&0\\0&1&0\\0&0&0\\0&0&1\\0&0&0\\0&0&0\\0&0&0\\0&0&0 \end{bmatrix}, \quad U_{SM,k+3} = \begin{bmatrix} 1&0&0&0\\0&1&0&0\\0&0&0&0\\0&0&1&0\\0&0&0&0\\0&0&0&0\\0&0&0&0\\0&0&0&0\\0&0&0&1\\0&0&0&0 \end{bmatrix} \tag{5.22}$$

The error covariance calculated at instant $(k + 2)$ is

$$U_{C01} = \begin{bmatrix} 1&0&0&0&0\\0&0&1&0&0\\0&0&0&0&1 \end{bmatrix}, \quad U_{C01}^{\mathrm{T}} P_{C,k+1} U_{C01} = \begin{bmatrix} p_{22}&0&p_{24}&0&p_{27}\\0&0&0&0&0\\p_{42}&0&p_{44}&0&p_{47}\\0&0&0&0&0\\p_{72}&0&p_{74}&0&p_{77} \end{bmatrix} \tag{5.23}$$

$$U_{MC,k+2} = \begin{bmatrix} 0&1&0&0&0\\0&0&1&0&0\\0&0&0&1&0 \end{bmatrix}, \quad P_{C,k+2} = \begin{bmatrix} p_{22}&0&p_{24}&0&p_{27}\\0&p_{33}&p_{34}&p_{36}&0\\p_{42}&p_{43}&p_{44}-\bar{p}_{44}&p_{46}&p_{47}\\0&p_{63}&p_{64}&p_{66}&0\\p_{72}&0&p_{74}&0&p_{77} \end{bmatrix} \tag{5.24}$$

$$U_{CS,k+2} = \begin{bmatrix} 0&1&0&0&0&0&0&0&0&0\\0&0&1&0&0&0&0&0&0&0\\0&0&0&1&0&0&0&0&0&0\\0&0&0&0&0&1&0&0&0&0\\0&0&0&0&0&0&1&0&0&0 \end{bmatrix} \tag{5.25}$$

## 5.5 Simulation results

### *5.5.1 Sampling of complex field with centralized AS algorithm using four robots along with partitioning of sampling area*

A complex fire field of size 124 $\times$ 90 pixels generated using a cellular automata (CA) simulation is to be reconstructed in parallel by sampling using four robots. An initial approximation is acquired by training the radial basis function (RBF) neural network with 30 neurons with spread factor of 25. The numerical covariance assumptions are

$$\begin{aligned} &P_{b0} = 5, \quad P_{a0} = 5, \quad P_{\sigma 0} = 10^{-4}, \quad P_{x00} = 5, \quad P_{y00} = 5 \\ &R = 0.01, \quad n = 15, \quad p = 5 \end{aligned} \tag{5.26}$$

Figure 5.9b and 5.9c, respectively, shows the initial field, and final estimated field after 172 samples. Figure 5.9e indicates the locations where the four robots took samples, represented by different symbols. It also shows the four regions formed using FCM algorithm and the Voronoi tessellation diagram, where different robots sample. Regions are updated after every 20 samples. The partitioning algorithm is implemented with *fcm* and *voronoi* functions of MATLAB. As shown in Figure 5.9d and 5.9g, the parametric error drops from 47.7 to 14.83, and the non-parametric error drops from 24.34 to 10.62 by taking 172 samples. The estimate of 120 parameters of 30 Gaussians is updated using the centralized AS algorithm.

The use of four robots instead of one for sampling also reduces the time for field reconstruction from 6.7 to 1.2 min (more than four times reduction). The reason can be understood intuitively because by sampling using four robots, not only the number of samples per robot reduces, but also navigation time reduces because of a smaller sampling area allocated to each robot.

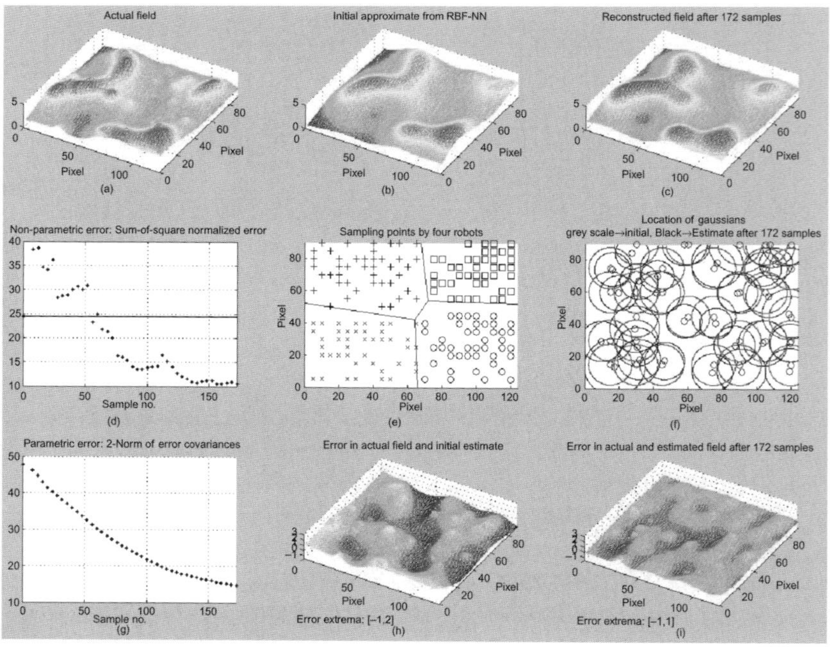

*Figure 5.9 (a–i) Sampling operation divided between four robots using GAS algorithm to reconstruct the field. The results plotted are after 172 samples*

Table 5.2 illustrates the number of computations and communications involved in sampling of complex field represented with 30 Gaussian RBF, using four robots, and considering a decentralized and distributed federated filter. Since the decentralized filter is full-order, it includes estimates of all 120 parameters. The number of computations performed on the fusion filter is same for decentralized and distributed filter. The combined computations for distributed filter are 675,840 compared with 8,631,840 for decentralized federated filter, e.g. approximately 12 times smaller. Similarly, the number of data communicated in case of distributed filter is 15 times smaller compared with the decentralized filter.

*Table 5.2 Comparison of decentralized and distributed federated filter for sampling of a complex field represented with 30 Gaussian RBF using four robots*

| | Computations | | | Communication |
|---|---|---|---|---|
| | **Robot** | **Fusion centre** | **Combined** | |
| Decentralized federated filter | $p = 16$, $m = 10$, $n = 120p$ $(2m^2n + 2mn^2 + m^3 + n^3)$ $= 2{,}041{,}000$ | $p = 16$, $n = 120$, $N = 4p(2n^2 + n^2N + nN)$ $= 467{,}840$ | 8,631,840 | $pN(n^2 + n)$ $= 929{,}280$ |
| Distributed filter | $p = 16$, $m = 10$, $n = 30p(2m^2n + 2mn^2 + m^3 + n^3)$ $= 52{,}000$ | $p = 16$, $n = 120$, $N = 4p(2n^2 + n^2N + nN)$ $= 467{,}840$ | 675,840 | $pN(n^2 + n)$ $= 59{,}520$ |

## 5.6 Experimental results

### *5.6.1 Sampling of linear colour field with centralized AS algorithm using two robots*

A linear RGB field represented by nine parameters was considered in the experiments of Chapter 3. The same colour field was reconstructed using two robots with an AS algorithm running on the central base station. A star network was formed, and the base station runs the AS algorithm and communicates with two Automation & Robotics Research Institute (ARRI)-Bots. The base station uses both robots to estimate the field parameters in half the time if data packet collisions are not encountered. If the sampling space is partitioned *a priori*, the results shown in Figure 5.10 (see p. C15) indicate that the colour field is again recovered with reasonable accuracy after 30 combined samples.

### *5.6.2 Sampling of complex fire field with centralized AS algorithm using two robots*

Experimental results are shown in Figure 5.11 for a complex fire field projected in an area of 124 × 90 square inches. Figure 5.11a shows the field that is generated

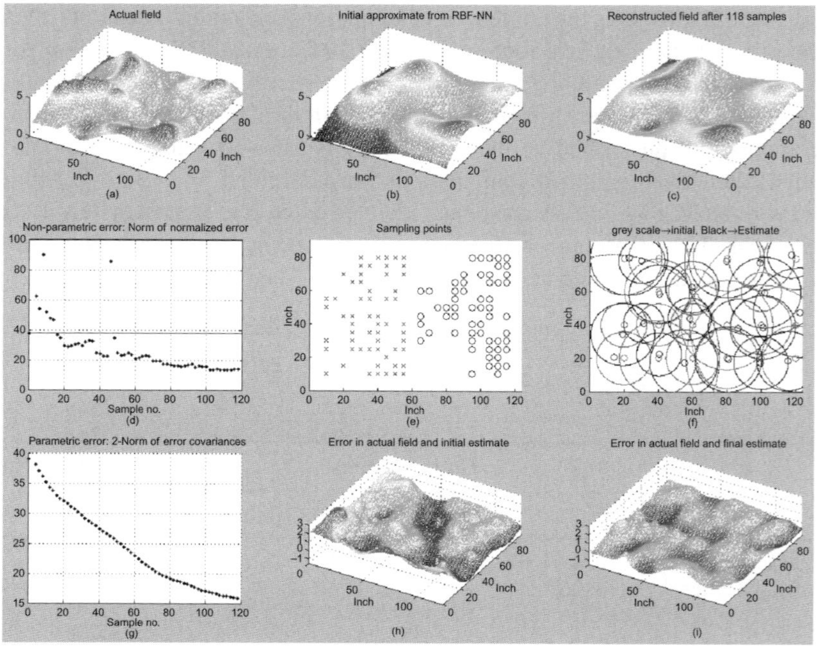


*Figure 5.11 (a–i) Experimental results for estimation of a complex field represented by 20-neuron RBF using two robots*

using the cellular automaton. A low-resolution grid size with $n = 20$ gives 27 ($3 \times 3$) samples for training the neural network. We use 20 neurons and a spread factor of 25 to approximate the field with 80 parameters ($20 \times 4$). Two robots sample the field in parallel using the GAS algorithm with grid size $p = 5$ and horizon size 3.

Figure 5.11b and 5.11c, respectively, shows the initial field and final estimated field after 118 samples. Figure 5.11e indicates the locations by '$\times$' and 'o' symbols where the two robots took samples. As shown in Figure 5.11d and 5.11g, the parametric error drops from 37.6 to 13.91, and the non-parametric error drops from 39.05 to 15.86 by taking 118 samples. This final non-parametric error is close to what we would get if we sample data from an actual field assuming a grid size of five (which uses 446 samples) to train the neural network. The numerical assumptions for uncertainties were

$$\begin{aligned} &P_{b0} = 5, \quad P_{a0} = 5, \quad P_{\sigma 0} = 10^{-4}, \quad P_{x00} = 5, \quad P_{y00} = 5 \\ &R = 0.01 \end{aligned} \tag{5.27}$$

The use of two robots instead of one for sampling also reduces the time for reconstruction from 6.7 to 2.6 min (which is less than half). This can be understood intuitively because not only the number of samples per robot reduces, but also navigation time reduces in a smaller sampling area.

## 5.7 Summary

In summary, in this chapter we propose a distributed federated filter, and a clustering scheme for the distribution of computational and communication load among $N$ robots performing sampling missions. Estimates of communication and computational load on the robots show that a dramatic reduction can be expected compared to a centralized sampling scheme, while a reduced sampling time in excess of $N$ times can be expected. Further simulations and experiments are needed to verify the efficiency and convergence properties of the distributed filter scheme we propose.

# Chapter 6
# Resource scheduling

In manufacturing systems, resources are usually application specific with slight flexibility of resource assignment to tasks, whereas in mobile sensor networks, the resources are heterogeneous and capable of performing diverse tasks. Hence, we have shared resources where multiple tasks contend for a single shared resource, or multiple resources contend to perform a single task. In the former case, we have shared resources, and in the latter, routing resources. The need then arises to suitably assign, dispatch and schedule resources in such a manner so as to avoid contention, or circular wait (CW) of resources leading to deadlock.

This chapter is organized into the following sections. Section 6.1 discusses the matrix-based discrete event controller (DEC), section 6.2 introduces deadlocks and presents the deadlock avoidance policy along with implementation on the wireless sensor network (WSN) test bed, section 6.3 discusses the issues of deadlock avoidance in the presence of routing resources and section 6.4 presents simulation results of deadlock avoidance both with and without routing resources. Experimental results of deadlock avoidance on the Automation & Robotics Research Institute (ARRI) WSN test bed are also presented in this section.

## 6.1 Matrix-based discrete event controller

A patented matrix formulation [107] is presented for modelling and analysis of complex interconnected discrete event (DE) systems needing dynamic online resource assignment in the presence of shared resources. The DEC is a hybrid system with logical and algebraic components that allows fast, direct design and reconfiguration of rule-based controllers [108]. The matrix approach provides a rigorous, yet intuitive mathematical framework to represent the dynamic evolution of DE systems through linguistic *if–then* rules:

*Rule i*: *If* <condition$^i$>, then <consequent$^i$>

The framework of the DEC is described that provides a rigorous simple representation of these linguistic rules. Let $r$ be the set of resources in the system (e.g. various mobile robots and unattended ground sensors (UGS)), $v$ be the set of tasks that the resources can perform (e.g. take a sensor reading, navigate to a commanded location along a desired path and retrieve/deploy UGS), $u$ be the set of inputs that trigger the system (e.g. detection of events such as chemical alert, intruder alert, etc., node failures), $y$ be the set of outputs indicating completed missions and $x$ be the logical state vector of rules of the DEC indicating the activated rules of the supervisory control policy.

The condition and consequent of each rule are segregated by the two sets of logical equations, one for checking the prior conditions leading to the activation of rule $i$ (matrix controller state equation) and one for determining the *a priori* consequent of rule $i$ (matrix controller output equation). The logical equations make use of matrix algebra for multiplications and additions with the element multiplications replaced by logical AND operations and the element additions replaced by logical OR operations. Logical negations are indicated by overbars.

The matrix controller state equation is

$$\overline{x} = F_v\overline{v} + F_r\overline{r} + F_u\overline{u} + F_{uc}\overline{u}_c \tag{6.1}$$

where $F_v$ is the task sequencing matrix, $F_r$ is the resource requirement matrix, $F_u$ is the input matrix, $F_{uc}$ is the conflict resolution matrix and $u_c$ is the conflict resolution vector. $u_c$ along with $F_{uc}$ is used to inhibit simultaneous activation of conflicting rules. The state of the DE system is maintained in the $x$, $v$, $r$ and $u$ vectors whose active (*true*) entries indicate the activated rules, the completed tasks, the available resources and the occurrence of events, respectively.

The task sequencing matrix $F_v$ has element $(i, j)$ set if the completion of task $v_j$ is an immediate prerequisite for the activation of logic state $x_i$. The resource requirement matrix $F_r$ has element $(i, j)$ set if the availability of resource $r_j$ is an immediate prerequisite for the activation of logic state $x_i$.

The matrix controller state equation (6.1) defines the prior conditions required for the activation of a rule, while the matrix controller output equations (6.2)–(6.4) define the *a priori* consequents of a rule.

The matrix controller output equations are

$$v_s = S_v x \tag{6.2}$$

$$r_s = S_r x \tag{6.3}$$

$$y = S_y x \tag{6.4}$$

where $S_v$ is the task start matrix having element $(i, j)$ set if logic state $x_j$ determines the activation of task $v_i$ and $S_r$ is the resource release matrix having element $(i, j)$ set if logic state $x_j$ determines the release of resource $r_i$.

Equations (6.2)–(6.4) represent the rule base of the supervisory control of the DE system. All the matrices are composed of sparse Boolean entries, so that real-time control of large interconnected DE systems with multiple missions and routes is computationally feasible.

The DEC has been used as a supervisory control in mobile sensor networks as detailed in Reference 109.

Figure 6.1 illustrates the DEC architecture.

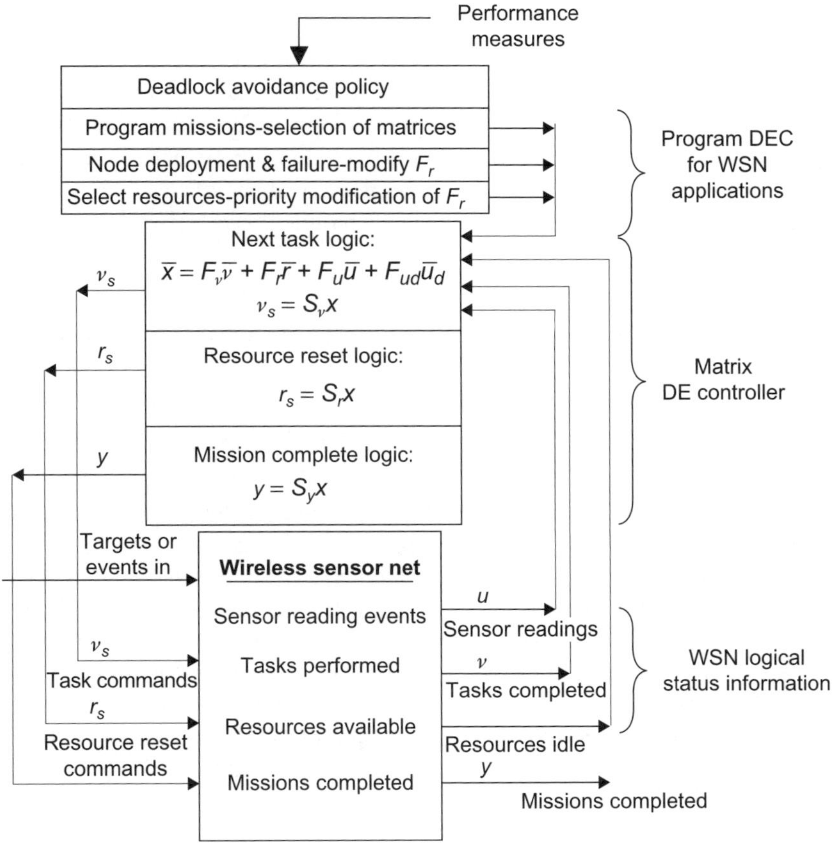


*Figure 6.1 Discrete event control architecture*

## 6.2 Deadlock avoidance

Deadlock research in computer systems has focused on four main areas. *Deadlock prevention* is involved with removing any possibility of system deadlocks; the result is often overconservative polices resulting in poor utilization of resources. *Deadlock detection* focuses on detecting imminent or current deadlocks and is required for deadlock recovery and avoidance strategies. *Deadlock recovery* methods are used to clear deadlocks once they occur, often by placing jobs in buffers or by completely flushing one or more of the deadlocked processes, resulting in lost work. In *deadlock avoidance*, the possibility of system deadlock is not totally removed, but whenever a deadlock is imminent, it is sidestepped by a real-time decision-making procedure [51]. In this book, we focus on deadlock avoidance.

### 6.2.1 Deadlock avoidance policy

Deadlock-free dispatching rules are derived by performing CW analysis in matrix form for possible deadlock situations. An analysis of deadlocks in manufacturing systems using the matrix-based DEC is presented in References 51 and 64. Deadlock avoidance algorithms have been implemented in robotic cells in manufacturing systems using the DEC [52,53]. Preliminary analysis of deadlock avoidance policies for shared resources in heterogeneous mobile sensor networks is presented in Reference 57.

For deadlocks, we consider the following assumptions:

- *No preemption*: No resource can be removed from a task until the completion of the task.
- *Mutual exclusion*: Every resource performs only one task at a given time.
- *Hold while waiting*: A process holds the resources allocated to it until it has all resources required to perform a job.

Under these assumptions, a necessary condition for deadlock to occur is the presence of a CW relation among the resources [51,53,62–64].

For any two resources $r_i$ and $r_j$, $r_i$ is said to wait for $r_j$, denoted by $r_i \rightarrow r_j$, if the availability of $r_j$ is an immediate requirement for the release of $r_i$. CWs among resources are a set of resources $r_a, r_b, \ldots, r_w$ whose wait relationships among them are $r_a \rightarrow r_b \rightarrow \cdots \rightarrow r_w$ and $r_w \rightarrow r_a$. To identify simple circular waits (sCWs), a wait relation digraph of resources needs to be constructed. The digraph of resources is easily obtained from the following matrix formulation:

$$W = (S_r F_r)^{\mathrm{T}} \tag{6.5}$$

where element $W_{ij}$ is set if $r_i \rightarrow r_j$ holds. sCWs are calculated from the digraph matrix using string algebra [52]. The sCWs do not represent all the CWs, and we require the CW matrix that is composed of all sCWs along with unions of non-disjoint sCW. The Gurel algorithm [62,63] is used to efficiently compute all CWs and composed CWs, $C_{out}$.

In order to implement efficient real-time deadlock avoidance policies, other relevant sets of task and resources from Petri net theory need to be defined. The term *token* is used to indicate a task in progress or an available resource and the term *transition* to indicate a rule of the supervisory controller.

A *siphon* is a set of tasks and resources that if token-free after firing of a certain transition will remain token-free under all subsequent transition firings. The critical siphon of a CW is the smallest siphon containing the CW. If the critical siphon ever becomes empty, all its resources are busy and can never become available again. The *siphon-task* set $J_s(C)$ is the set of tasks of the critical siphon. The *critical subsystem* $J_o(C)$ is a set of tasks of the CW, $J(C)$, which do not add a token to the CW.

A deadlock condition occurs if and only if there is an empty CW, which corresponds to an empty critical siphon, or equivalently to a condition where all tasks of the CW belong to the critical subsystem.

Thus, in order to perform deadlock analysis, we need matrix computation tools to determine the siphon-task sets $J_s(C)$ and the critical subsystems $J_o(C)$ of every CW $C$. Since the deadlock conditions are dependent on the number of tokens in these sets, we need to calculate the set of transitions (input and output transitions), which when fired, add or subtract tokens from the CWs.

The input and output transitions of a CW are calculated as

$$\begin{aligned} {}_dC &= C_{out}S_r \\ C_d &= C_{out}F_r^{\mathrm{T}} \end{aligned} \tag{6.6}$$

The adding and clearing transitions are calculated as

$$\begin{aligned} T_p &= {}_dC - ({}_dC \wedge C_d) \\ T_m &= C_d - (C_d \wedge {}_dC) \end{aligned} \tag{6.7}$$

where the $\wedge$ operator represents logical AND.

These sets of transitions are important in keeping track, in real time, of the available resources inside every CW, and hence in determining the status of tasks and resources inside the critical siphon. The task set, siphon-task set and the critical subsystem of a CW are calculated as

$$\begin{aligned} J(C) &= {}_dCF_v = C_dS_v{}^{\mathrm{T}} \\ J_s(C) &= T_pF_v \\ J_o(C) &= J(C) \wedge (C_dF_v) \end{aligned} \tag{6.8}$$

The critical subsystem is the set of tasks that require one of the resources from the CW for execution. Therefore, the activation of all tasks in the critical subsystem will make the CW empty and lead to a deadlock condition. A simple deadlock avoidance strategy consists of keeping the number of activated tasks of a critical subsystem lesser than the number of resources in the corresponding CW. This is the MAXWIP policy:

$$m(J_o(C_i)) < m_o(C_i) \tag{6.9}$$

The described deadlock avoidance policy has been implemented on the mobile WSN test bed and complex interconnected missions executed in a smooth, deadlock-free manner.

## 6.3 Routing resources

Resources in mobile sensor networks are heterogeneous in nature and capable of performing several diverse tasks unlike those in manufacturing systems where resources are usually dedicated to a specific task. Thus, for every task awaiting execution, multiple resources are available for being scheduled and the task can request the best (cost/time/energy efficient) resource that can take the task to completion in an optimal manner. Hence, we need a formal mathematical model for deadlock-free dynamic resource scheduling where routing as well as dispatching decisions need to be made.

Deadlock analysis in the presence of routing choices is of exponential complexity, and deadlock avoidance constraints are rendered computationally intractable [65].

Very little research exists in the field of mobile sensor networks where resources are scheduled dynamically for task execution in a deadlock-free manner. This book presents a mathematical formulation for dynamic resource scheduling using the matrix-based DEC. A deadlock avoidance algorithm is developed and simulated for task sequencing in the presence of routing choices.

### 6.3.1 DEC representation for routing

The matrix-based DEC presented in an earlier section (see (6.1)–(6.4)) is flexible and can be easily used to implement the scenario of task sequencing where predetermined resource assignments do not exist and dynamic online resource scheduling needs to be performed for allocating a resource to a particular task.

In the presence of shared resources, or in the case of online resource assignment (routing resources), simultaneous activation of conflicting rules arises. The conflict resolution matrix, $F_{uc}$, in (6.1) is used to resolve conflicts. In an earlier work on deadlock avoidance in the presence of shared resources in mobile sensor networks [57], the conflict resolution policy had to handle conflicts of shared resources, i.e. conflicts derived by the simultaneous activation of rules that start different tasks requiring the same resource. However, in the current scenario, with no predetermined resource assignments, conflicts of pseudo-shared resources and conflicts of routing resources arise.

Conflicts of pseudo-shared resources arise when simultaneous activation of rules that start the same job using the same resource but having different consequents (such as releasing different resources) occurs. This happens when multiple paths through the sequence join, for instance, logic transitions $t_4$ and $t_5$ in Figure 6.2 where the resource $B_2$ acts like a shared resource but starts the same task.

Conflicts of routing resources arise when simultaneous activation of rules that start the same job using different resources (different routes) occurs. This happens when a single route splits, where the same job can be assigned from among a set of resources, for instance, logic transitions $t_2$ and $t_3$ in Figure 6.2 where resources $R_1$ and $R_2$ can perform the same job.

A novel augmented conflict resolution matrix is proposed:

$$F_{uc} = F_{uc\text{-}shared} || F_{uc\text{-}pseudo\text{-}shared} || F_{uc\text{-}routing} \tag{6.10}$$

such that element $(i, j)$ is set if completion of conflict-arising task $v_j$ is an immediate prerequisite for the activation of logic state $x_i$. Then setting the element $j$ in the conflict resolution vector $u_c$ determines the inhibition of logic state $x_i$ (rule $i$ cannot be fired). Thus, depending on the way one selects the conflict resolution strategy to generate $u_c$, different dispatching strategies can be selected to avoid resource conflicts due to shared resources, pseudo-shared resources and routing resources.

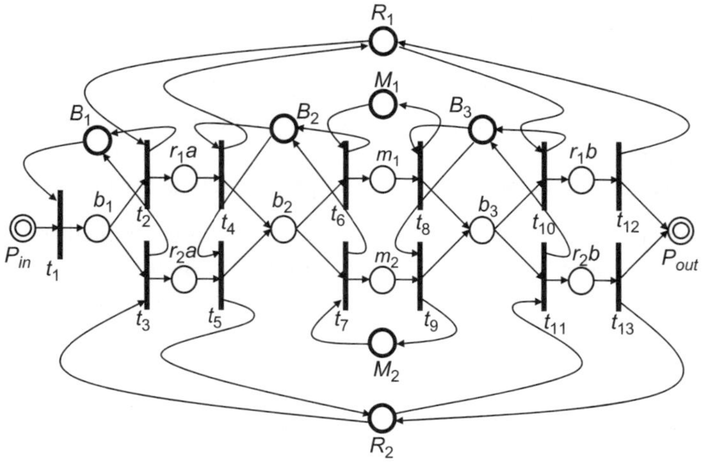


*Figure 6.2 Sample Petri net with routing resources*

Figure 6.3 depicts the construction of the augmented conflict resolution matrix from elements of the resource requirements, and the resource release matrices.

## 6.3.2 Deadlock avoidance policy for flexible routing systems

An analysis of deadlock structures using matrices for reentrant flow lines with routing in a flexible manufacturing system is performed in Reference 67. However, these mathematical constructions give us a method of detecting active CWs only, since in systems with routing choices, an active CW does not always lead to a deadlock. In this section, we present mathematical formulations for detecting when an active CW could become a potential deadlock, and we also present a computationally feasible deadlock avoidance algorithm.

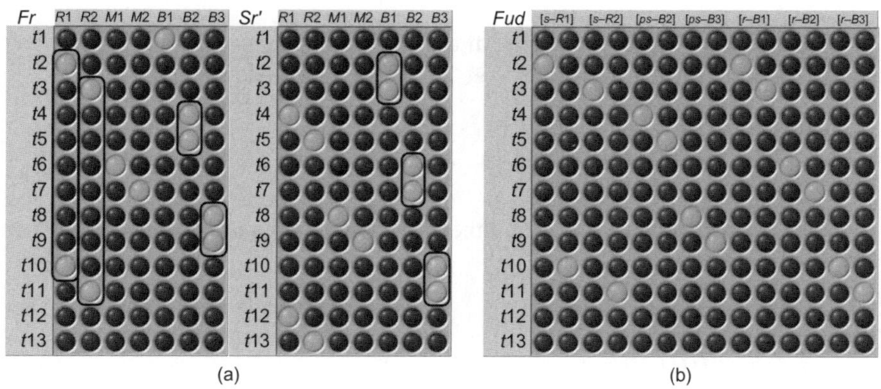


*Figure 6.3 Augmented conflict resolution matrix formulation. (a) $F_r$ and $S_r^T$ matrices for the sample Petri net. Shared, pseudo-shared and routing resources are highlighted. (b) Augmented conflict resolution matrix*

In addition to assumptions for deadlock that were made in section 6.2.1, we have the following non-restrictive capabilities:

- Some tasks have the option of being executed by a resource from a set of resources (routing resources), and each resource might be used for different tasks (shared resources).
- Task routings are deterministic and are provided by a dynamic controller.

Mathematical constructions defined in section 6.2.1 are modified as suggested in Reference 67 for adapting the deadlock analysis for the more complex case of routing. Because of the diversity of loop paths that a set of resources contained in an sCW might have (section 6.2.1, (6.5)), we need to identify not only the resources that compose the sCW, but also the transitions that link them. This will give us specific information needed to locate siphons and certain critical subsystems needed for construction of the deadlock policy. A general digraph matrix is used as follows:

$$W = \begin{bmatrix} 0_r & S_r \\ F_r & 0_t \end{bmatrix} \tag{6.11}$$

where $0_r$, $0_t$ are $r \times r$, $t \times t$ zero matrices respectively, with $r$ being the number of resources and $t$ being the number of transitions.

Using the general digraph matrix, $W$, we get both sCW of resources and sCW of transitions:

$$C_w = [\,C_{wr*} \quad C_{wt*}\,] \tag{6.12}$$

and using the Gurel algorithm, we obtain the matrix $G$ of composed CWs to get all the CWs of resources and transitions:

$$\begin{aligned} CW_r &= G^{\mathrm{T}} C_{wr*} \\ CW_t &= G^{\mathrm{T}} C_{wt*} \end{aligned} \tag{6.13}$$

The input and output transitions of a CW are still calculated as (6.6); however, the adding and clearing transitions (see (6.7)) are modified as

$$\begin{aligned} T_p &= {}_dC - ({}_dC \wedge CW_t) \\ T_m &= C_d - (C_d \wedge CW_t) \end{aligned} \tag{6.14}$$

The task set, siphon-task set and the critical subsystem of a CW (see (6.8)) are also modified as

$$\begin{aligned} J_C &= {}_dCF_v = C_d S_v^{\mathrm{T}} \\ J_s(C) &= J(C) \wedge (\overline{C_d F v}) \\ J_o(C) &= CW_t F_v \end{aligned} \tag{6.15}$$

With these formulations (see (6.11)–(6.15)), we can detect only active CWs but not when an active CW progresses to a potential deadlock. In a system with routing choices, the system can exist in an active CW and still not cause a deadlock. In this book, we introduce the concept of exit policies, exit transitions and exit CW to try to detect when an active CW may progress to a potential deadlock.

The exit transition matrix, *XT*, is introduced, which is the set of transitions that when fired would introduce a token into an empty CW. Such transitions only exist in systems with routing choices. The exit transition matrix is given as

$$XT = J_o(CW) \bullet -CW_t \tag{6.16}$$

The exit CW matrix, $CW_X$, defines the set of CWs into which an active CW could exit on firing an exit transition. On introducing a token into an active CW, an exit transition could cause another CW to become empty. The $CW_X$ matrix defines the set of such CWs for a particular CW. The exit CW matrix is given as

$$CW_X = \{CW_t : XT \in CW_t\} \tag{6.17}$$

The proposed deadlock avoidance policy allows a CW to progress into an active CW provided there exists at least one or more exit transitions that when fired would clear the current CW without transiting another CW into an active CW. Hence, one or more free exit transitions need to exist, expressed mathematically as

$$n(CW_{active}) \in CW_X > 0 \tag{6.18}$$

Thus, our deadlock avoidance policy allows an empty CW to form provided a free exit transition exists. This is a computationally feasible solution for deadlock-free dispatching in the presence of routing choices. Figure 6.4 illustrates the various constructs discussed in this section for the four sCWs of the sample Petri net in Figure 6.2.

A more complete analysis of deadlock avoidance in the presence of routing resources is continued in Reference 110.

## 6.4 Simulation and experimental results

In our motivating scenario, a mobile WSN is in charge of monitoring a warehouse where dangerous chemicals are handled. Based on a fair knowledge of the environment and the possible operating conditions, it is possible to come up with predefined sequences of tasks that the robots have to accomplish in response to external threats or programmed events.

A virtual WSN test bed has been created to illustrate various mobile robot movements as the WSN topology reconfigures to handle various missions. Figure 2.10b illustrates a view of the virtual WSN test bed in the initial network configuration.

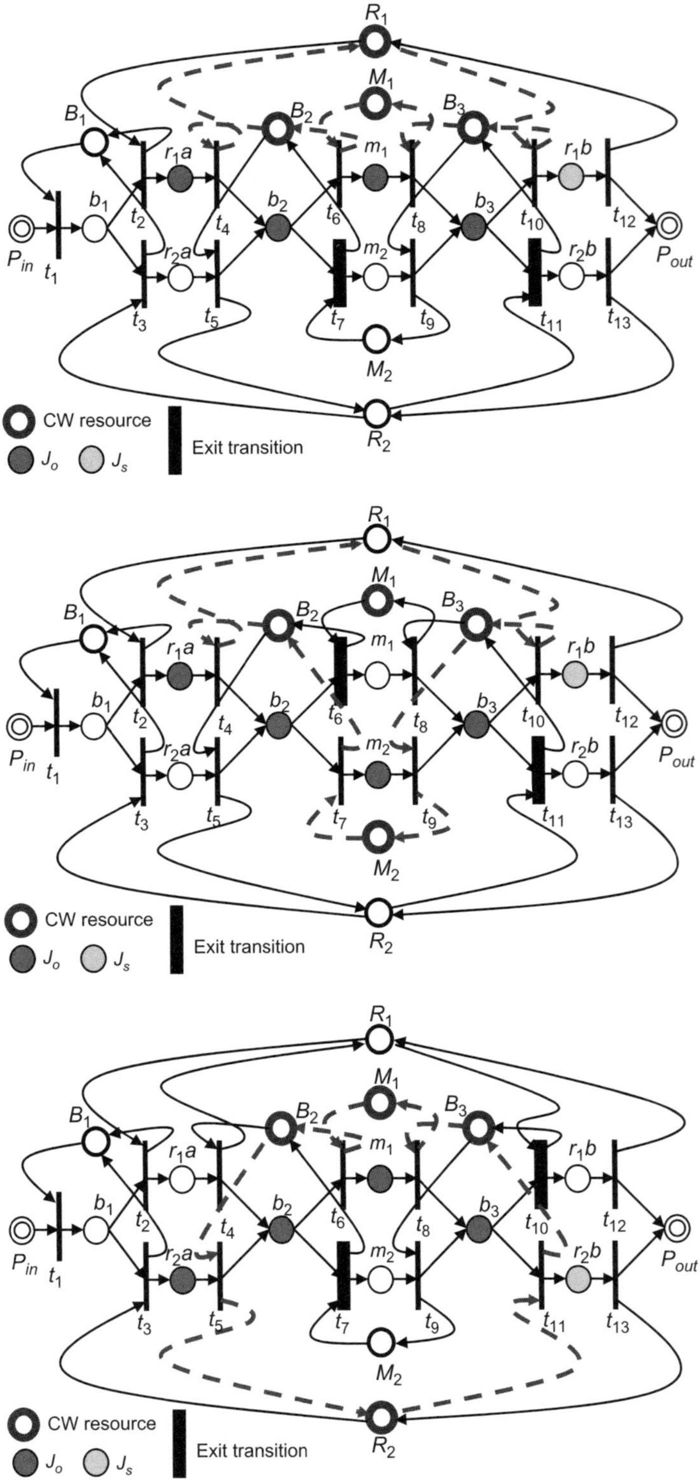


*Figure 6.4 Four simple CWs and their corresponding $CW_r$, $CW_t$, $J_s$, $J_o$ and XT*

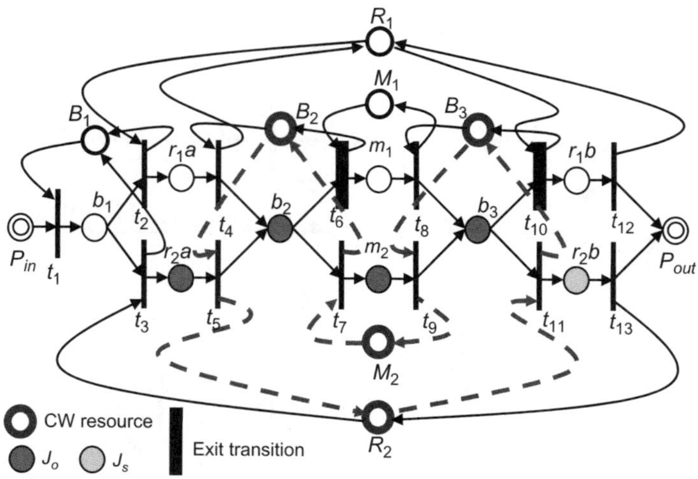


*Figure 6.4 Continued*

## 6.4.1 Simulation and experimental results for deadlock avoidance

The results presented in this chapter have been obtained using MATLAB and LabVIEW programming environments. MATLAB has been used for initial simulation of the missions, followed by a LabVIEW implementation of the missions with *simulated resources*. With satisfactory performance of the deadlock resolution algorithm, the simulated resources were replaced by *real resources* and the missions were actually implemented in our laboratory. Thus, the same code has been used for simulation and real implementation of all missions. The similarity and fidelity of the dispatching sequences in both the simulation and experimental phases were indeed satisfactory. This is a key result since it shows that the proposed DEC allows one to perform a 'simulate and experiment' approach for an MSN, with noticeable benefits in terms of cost, time and performance.

Three separate missions have been implemented to illustrate the proposed control architecture. The first mission aims at patrolling the warehouse, the second mission aims at recharging the UGS batteries and the third mission aims at transporting dangerous chemicals to a predefined location. The triggering events of the three missions are an intruder alert, a battery low alert and a prescheduled instant of time, respectively. Once the alert is received, the network physically reconfigures its topology to react to the alert. For the sake of completeness, the Petri net representation of missions 1, 2 and 3 is illustrated in Figure 6.5.

The procedure for implementing the proposed control policy consists of three different steps.

First, we define the vector of resources $r$ of the system. In this example, we have two robots $R$ and six stationary sensors $M$. The resource vector is $r = [R_1, R_2, M_1, M_2, M_3, M_4, M_5, M_6]$.

Then for each mission $i$, we define the vector of inputs $u^i$, of outputs $y^i$ and of tasks $v^i$, and the task sequence of each mission (see Tables 6.1, 6.3 and 6.5 for missions 1, 2 and 3, respectively), and write down the *if–then* rules representing the

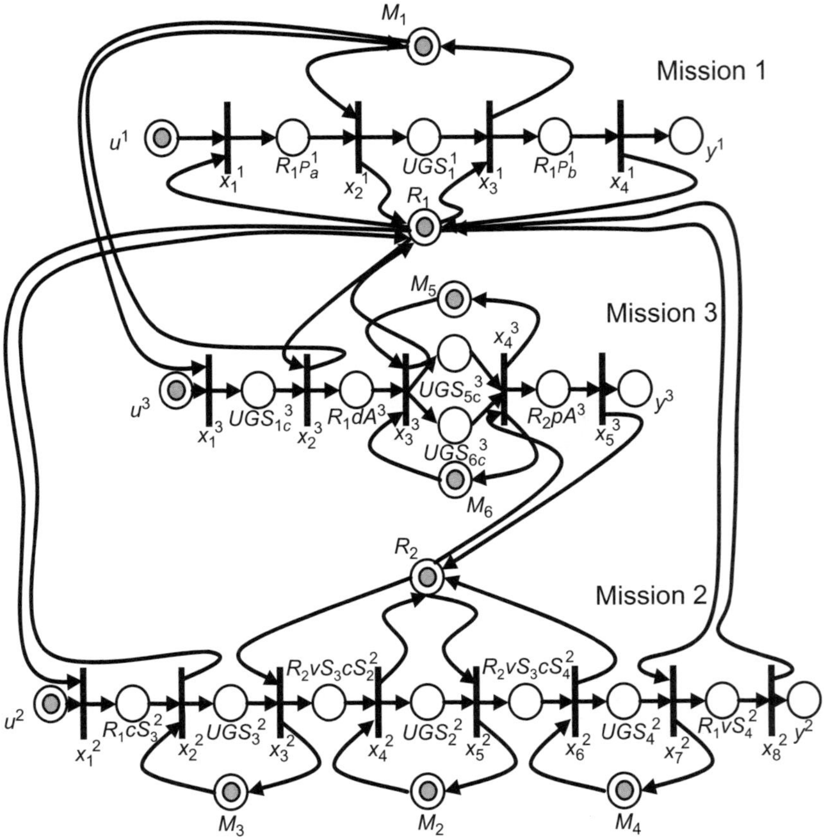


*Figure 6.5* *Petri net representation of mission 1 (patrol and sensing of the warehouse), mission 2 (charging of the UGS) and mission 3 (transportation of dangerous chemicals)*

supervisory coordination strategy to sequence the programmed missions (see Tables 6.2, 6.4 and 6.6 for missions 1, 2 and 3, respectively).

Finally, we translate the linguistic description of the coordination rules into a more convenient matrix representation, suitable for mathematical analysis and computer implementation. As an example, matrices $F_v^{\,1}$ and $F_r^{\,1}$ relative to mission 1 are reported (see Figure 6.6).

The CW matrix $C_{out}$ and the critical subsystem matrix $J_o$ are fairly complex for these interconnected missions. Figure 6.7 illustrates these matrices.

Experiment 1: *Deadlocks caused by multiple activation of the same mission*

Suppose that two different intrusion threats are detected and that, in the meantime, two sensors launch a low-battery alert. In this case, both missions 1 and 2 are triggered two times and deadlock conditions occur (Figure 6.8). In particular, we get two empty CWs: $R_1 \rightarrow M_1$ and $R_2 \rightarrow M_2$.

As an example, let us analyse the first CW. As can be seen from the first row of the CW matrix (Figure 6.7a), resources $R_1$ and $M_1$ form CW 1. And from the first

*Table 6.1 Mission 1 – task sequence*

| Mission 1 | Notation | Description |
|---|---|---|
| *Input* | $U^1$ | Intruder alert from any UGS |
| *Task 1* | $R_1P_a{}^1$ | (i) $R_1$ navigates to $M_2$ |
| | | (ii) $R_1$ takes measurement at $M_2$ |
| | | (iii) $R_1$ navigates from $M_2$ to $M_1$ |
| | | (iv) $R_1$ takes measurement at $M_1$ |
| *Task 2* | $UGS_1^1$ | (i) $M_1$ takes measurement |
| *Task 3* | $R_1P_b{}^1$ | (i) $R_1$ navigates to $M_1$ |
| | | (ii) $R_1$ takes measurement at $M_1$ |
| | | (iii) $R_1$ navigates from $M_1$ to $M_3$ |
| | | (iv) $R_1$ takes measurement at $M_3$ |
| *Output* | $y^1$ | (i) Patrol and sensing of warehouse |

*Table 6.2 Mission 1 – rule base*

| | **Mission 1 – operation sequence** |
|---|---|
| *Rule 1*, $x_1^1$ | If $u^1$ *occurs* and $R_1$ *available*, then *start* $R_1P_a{}^1$ |
| *Rule 2*, $x_2^1$ | If $R_1P_a{}^1$ *completed* and $M_1$ *available*, then *release* $R_1$ and *start* $UGS_1^1$ |
| *Rule 3*, $x_3^1$ | If $UGS_1^1$ *completed* and $R_1$ *available*, then *release* $M_1$ and start $R_1P_b{}^1$ |
| *Rule 4*, $x_4^1$ | If $R_1P_b{}^1$ *completed*, then *release* $R_1$ and *terminate* mission 1 by producing output $y^1$ |

*Table 6.3 Mission 2 – task sequence*

| Mission 2 | Notation | Description |
|---|---|---|
| *Input* | $U^2$ | Low-battery warning from UGS |
| *Task 1* | $R_1cS_3^2$ | (i) $R_1$ navigates to $M_3$ |
| | | (ii) $R_1$ charges $M_3$ |
| *Task 2* | $UGS_3^2$ | (i) $M_3$ takes measurement |
| *Task 3* | $R_2vS_3cS_2^2$ | (i) $R_2$ navigates to $M_3$ |
| | | (ii) $R_2$ takes measurement and verifies $M_3$ charge |
| | | (iii) $R_2$ navigates from $M_3$ to $M_2$ |
| | | (iv) $R_2$ charges $M_2$ |
| *Task 4* | $UGS_2^2$ | (i) $M_2$ takes measurement |
| *Task 5* | $R_2vS_2cS_4^2$ | (i) $R_2$ navigates to $M_2$ |
| | | (ii) $R_2$ takes measurement and verifies $M_2$ charge |
| | | (iii) $R_2$ navigates from $M_2$ to $M_4$ |
| | | (iv) $R_2$ charges $M_4$ |
| *Task 6* | $UGS_4^2$ | (i) $M_4$ takes measurement |
| *Task 7* | $R_1vS_4^2$ | (i) $R_1$ navigates to $M_4$ |
| | | (ii) $R_1$ takes measurement and verifies $M_4$ charge |
| *Output* | $y^2$ | (i) Charging of UGS |

Table 6.4 Mission 2 – rule base

| Mission 2 – operation sequence | |
|---|---|
| *Rule 1*, $x_1^2$ | If $u^2$ *occurs* and $R_1$ *available*, then *start* $R_1cS_3^2$ |
| *Rule 2*, $x_2^2$ | If $R_1cS_3^2$ *completed* and $M_3$ *available*, then *release* $R_1$ and *start* $UGS_3^2$ |
| *Rule 3*, $x_3^2$ | If $UGS_3^2$ *completed* and $R_2$ *available*, then *release* $M_3$ and *start* $R_2vS_3cS_2^2$ |
| *Rule 4*, $x_4^2$ | If $R_2vS_3cS_2^2$ *completed* and $M_2$ *available*, then *release* $R_2$ and *start* $UGS_2^2$ |
| *Rule 5*, $x_5^2$ | If $UGS_2^2$ *completed* and $R_2$ *available*, then *release* $M_2$ and *start* $R_2vS_2cS_4^2$ |
| *Rule 6*, $x_6^2$ | If $R_2vS_2cS_4^2$ *completed* and $M_4$ *available*, then *release* $R_2$ and *start* $UGS_4^2$ |
| *Rule 7*, $x_7^2$ | If $UGS_4^2$ *completed* and $R_1$ *available*, then *release* $M_4$ and *start* $R_1vS_4^2$ |
| *Rule 8*, $x_8^2$ | If $R_1vS_4^2$ *completed*, then *release* $R_1$ and *terminate* mission 2 by producing output $y^2$ |

Table 6.5 Mission 3 – task sequence

| Mission 1 | Notation | Description |
|---|---|---|
| *Input* | $u^3$ | 40 min has elapsed |
| *Task 1* | $UGS_{1c}^3$ | $M_1$ takes measurement |
| *Task 2* | $R_1dA^3$ | $R_1$ picks up dangerous cargo and drops off at temporary storage location A |
| *Task 3* | $UGS_{5c}^3 UGS_{6c}^3$ | $M_5$ and $M_6$ take measurements |
| *Task 4* | $R_2pA$ | $R_2$ picks up cargo from A and transports to location B along path decided by readings from $M_5$ and $M_6$ |
| *Output* | $y^3$ | *Dangerous cargo transported* |

Table 6.6 Mission 3 – rule base

| Mission 3 – operation sequence | |
|---|---|
| *Rule 1*, $x_1^3$ | If $u^3$ *occurs* and $M_1$ *available*, then *start* $UGS_{1c}^3$ |
| *Rule 2*, $x_2^3$ | If $UGS_{1c}^3$ *completed* and $R_1$ *available*, then *release* $M_1$ and *start* $R_1dA_3$ |
| *Rule 3*, $x_3^3$ | If $R_1dA_3$ *completed* and $M_5$ and $M_6$ *available*, then *release* $R_1$ and *start* $UGS_{5c}^3$ and $UGS_{6c}^3$ |
| *Rule 4*, $x_4^3$ | If $UGS_{5c}^3$ and $UGS_{6c}^3$ *completed* and $R_2$ *available*, then *release* $M_5$ and $M_6$ and *start* $R_2pA$ |
| *Rule 5*, $x_5^3$ | If $R_2pA$, then *release* $R_2$ and *terminate* mission 3 by producing output $y^3$ |

row of the critical subsystem matrix (Figure 6.7b), tasks $R_1P_a$ and $UGS_1$ form critical subsystem 1. When both these tasks are simultaneously in progress, CW 1 becomes empty and a deadlock occurs.

$$F_v^1 = \begin{array}{c} \begin{array}{ccc} R_1P_a & UGS_1 & R_1P_b \end{array} \\ \left[\begin{array}{ccc} 0 & 0 & 0 \\ 1 & 0 & 0 \\ 0 & 1 & 0 \\ 0 & 0 & 1 \end{array}\right] \end{array}, \quad F_r^1 = \begin{array}{c} \begin{array}{cccccccc} M_1 & M_2 & M_3 & M_4 & M_5 & M_6 & R_1 & R_2 \end{array} \\ \left[\begin{array}{cccccccc} 0 & 0 & 0 & 0 & 0 & 0 & 1 & 0 \\ 1 & 0 & 0 & 0 & 0 & 0 & 0 & 0 \\ 0 & 0 & 0 & 0 & 0 & 0 & 1 & 0 \\ 0 & 0 & 0 & 0 & 0 & 0 & 0 & 0 \end{array}\right] \end{array}$$

*Figure 6.6 Mission 1 job sequencing matrix ($F_v^1$) and resource requirement matrix ($F_r^1$)*

$$C_{out} = \begin{array}{c} \begin{array}{cccccccc} M_1 & M_2 & M_3 & M_4 & M_5 & M_6 & R_1 & R_2 \end{array} \\ \left[\begin{array}{cccccccc} \mathbf{1} & \mathbf{0} & \mathbf{0} & \mathbf{0} & \mathbf{0} & \mathbf{0} & \mathbf{1} & \mathbf{0} \\ \mathbf{0} & \mathbf{1} & \mathbf{0} & \mathbf{0} & \mathbf{0} & \mathbf{0} & \mathbf{0} & \mathbf{1} \\ & & & \vdots & & & & \\ \mathbf{0} & \mathbf{1} & \mathbf{1} & \mathbf{1} & \mathbf{1} & \mathbf{1} & \mathbf{1} & \mathbf{1} \\ \mathbf{1} & \mathbf{1} & \mathbf{1} & \mathbf{1} & \mathbf{1} & \mathbf{1} & \mathbf{1} & \mathbf{1} \end{array}\right]_{[30x8]} \end{array}$$

(a)

$$J_o = \begin{array}{c} \begin{array}{ccccccccc} R_1P_a & UGS_1 & R_1P_b & R_1cS_3 & & R_1dA & UGS_{5c} & UGS_{6c} & R_2pA \end{array} \\ \left[\begin{array}{ccccccccc} \mathbf{1} & \mathbf{1} & \mathbf{0} & \mathbf{0} & \ldots & \mathbf{0} & \mathbf{0} & \mathbf{0} & \mathbf{0} \\ \mathbf{0} & \mathbf{0} & \mathbf{0} & \mathbf{0} & \ldots & \mathbf{0} & \mathbf{0} & \mathbf{0} & \mathbf{0} \\ & & & & \vdots & & & & \\ \mathbf{0} & \mathbf{0} & \mathbf{0} & \mathbf{1} & \ldots & \mathbf{1} & \mathbf{1} & \mathbf{1} & \mathbf{0} \\ \mathbf{1} & \mathbf{1} & \mathbf{0} & \mathbf{1} & \ldots & \mathbf{1} & \mathbf{1} & \mathbf{1} & \mathbf{0} \end{array}\right]_{[30x15]} \end{array}$$

(b)

*Figure 6.7 (a) Circular wait matrix ($C_{out}$) and (b) critical subsystem matrix ($J_o$)*

In Figure 6.8a, the MATLAB simulation illustrates the triggering of mission 1 at time instants 14 and 19 for two different intrusion alerts. When the mission is triggered, task $R_1P_a$ starts and runs to completion. Then task $UGS_1^1$ starts. At around this time, mission 1 is triggered again and another instance of task $R_1P_a$ starts simultaneously. Now resources $R_1$ and $M_1$ are consumed. When task $UGS_1^1$ completes, $R_1P_b$ has to be fired and this requires resource $R_1$, which is already consumed. Also, when $R_1P_a$ completes, resource $M_1$ is needed for the task $UGS_1^1$ to fire. Since $M_1$ is being used, we have a cyclic wait of resources that leads to a deadlock situation. As shown in Figure 6.8b (path sequences $p_1, \ldots, p_5$ and $q_1, \ldots, q_4$ describe the motions of robots 1 and 2, respectively), robots 1 and 2 keep on wandering in the warehouse without accomplishing mission 1. Same considerations hold for mission 2 with the CW $R_2 \rightarrow M_2$.

Figure 6.9 illustrates the same case when the MAXWIP algorithm is applied. As can be seen, both missions successfully run to completion two times. In particular, for mission 1, in order to avoid deadlocks, the dispatching policy inhibits rule 1 $x_1^1$ when $UGS_1^1$ is in progress by conveniently updating the conflict resolution vector $u_d$. Similar considerations hold for mission 2. The controller inhibits $R_2$ from performing task $R_2vS_3cS_2^2$ as long as task $UGS_2^2$ is in progress. In this way, missions 1 and 2 are executed two consecutive times (Figure 6.9b and 6.9c), and the network successfully reacts to multiple intrusions and low-battery alarms.

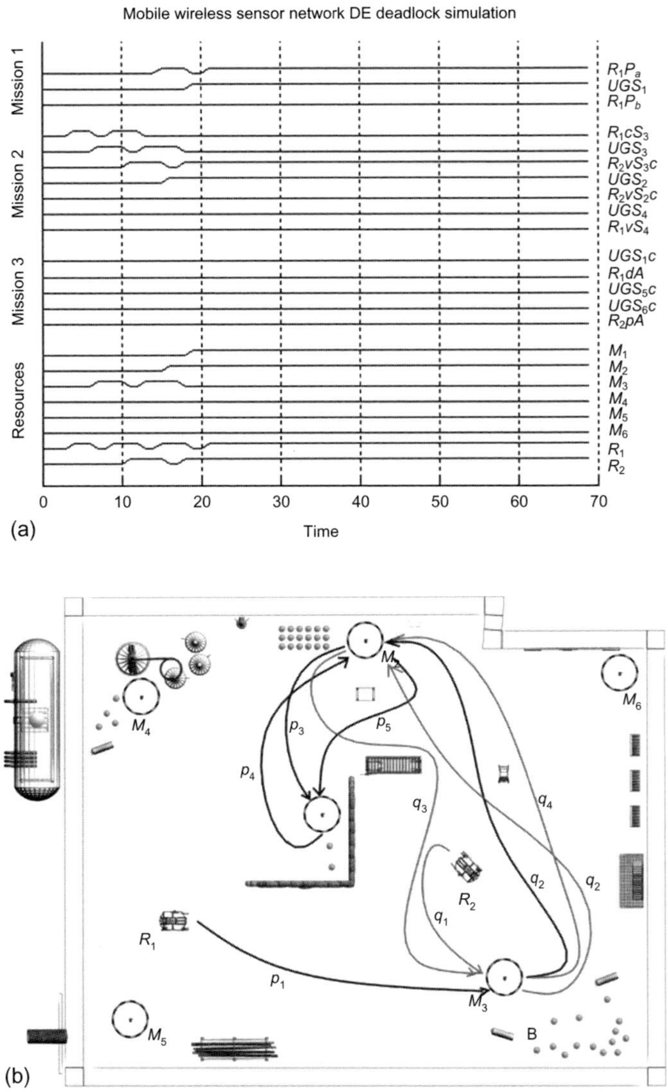


*Figure 6.8 Missions 1 and 2 deadlock: (a) event time traces and (b) top view of robots' paths (darker paths are $R_1$ paths; lighter paths are $R_2$ paths)*

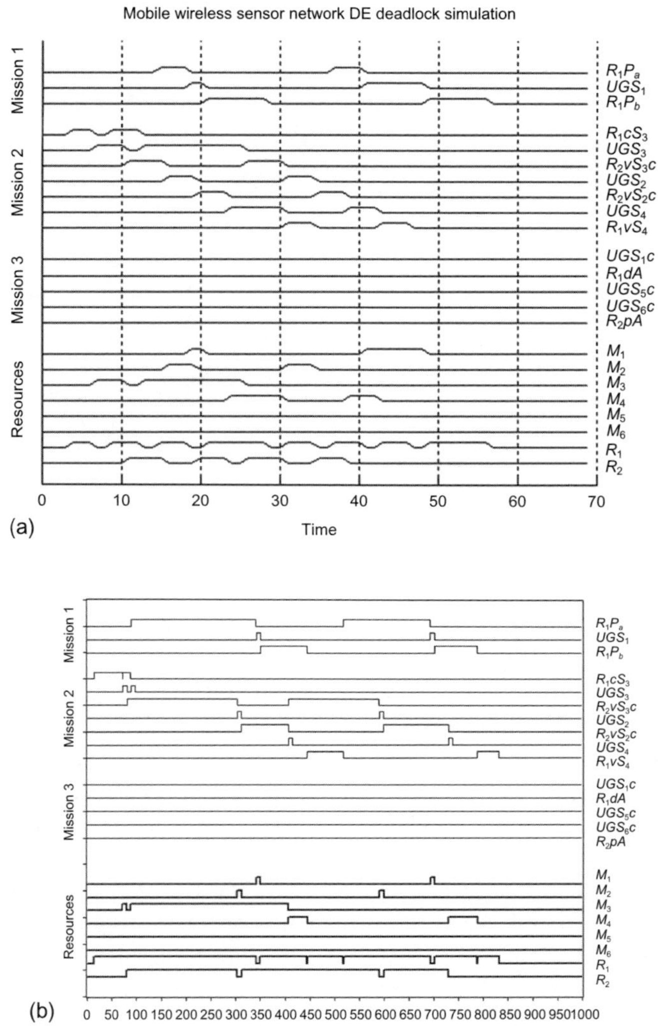


*Figure 6.9 Missions 1 and 2 deadlock avoidance: (a and b) event time traces – MATLAB simulation and LabVIEW implementation and (c) top view of robots' paths (darker paths are $R_1$ paths; lighter paths are $R_2$ paths)*

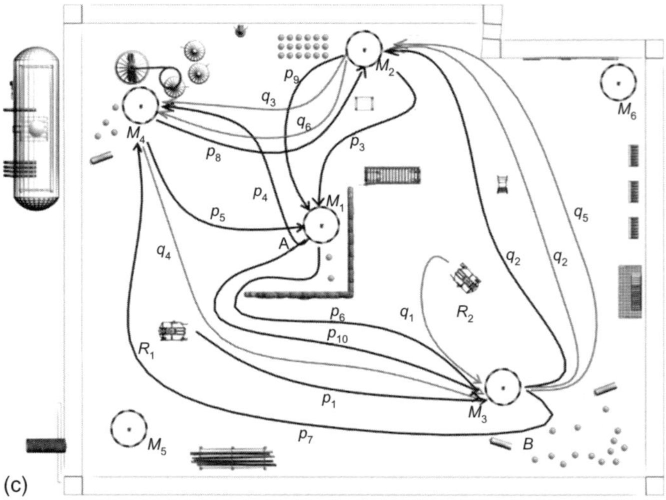


*Figure 6.9 Continued*

A comparison of Figure 6.9a and 6.9b shows that the task sequences in both simulation and experiment are the same (the different durations of the tasks in the two cases are not related to the behaviour of the DEC but to real-world navigation of the robots and are not therefore relevant to our discussion).

Experiment 2: *Deadlock caused by simultaneous activation of multiple missions*

As seen in the previous experiment, deadlock can occur due to multiple triggering of a certain mission. Deadlocks can also arise when two or more missions run in parallel and share a CW. This scenario is illustrated in Figure 6.10 wherein missions 1 and 3 execute in parallel and share the CW $R_1 \rightarrow M_1$. As shown in Figure 6.10a, when task $R_1P_a$ from mission 1 and task $UGS_{1c}$ from mission 3 execute in parallel, the CW $R_1 \rightarrow M_1$ gets empty and a deadlock occurs.

Figure 6.11a and 6.11b illustrates the simulation and the experimental results for the same scenario with the deadlock resolution algorithm applied. With no deadlock resolution, robot $R_1$ performs task $R_1P_a{}^1$ first and navigates to sensors $M_2$ and $M_1$ before getting stuck in a deadlock (Figure 6.10b). With the deadlock resolution algorithm, task $R_1P_a$ is inhibited until $UGS_{1c}$ is completed. $R_1$ performs task $R_1dA^3$ first and navigates to location A before successfully completing the assigned missions (Figure 6.11c).

### 6.4.2 Simulation results for routing

The sample Petri net considered in Figure 6.2 is simulated. This has multiple routing choices to be made, and the proposed augmented $F_{ud}$ matrix ensures that the conflicts that arise due to shared, pseudo-shared and routing resources are resolved. Initial simulations for deadlock consist of disabling all routing choices and triggering the mission multiple times (ten times); this is the case of simple

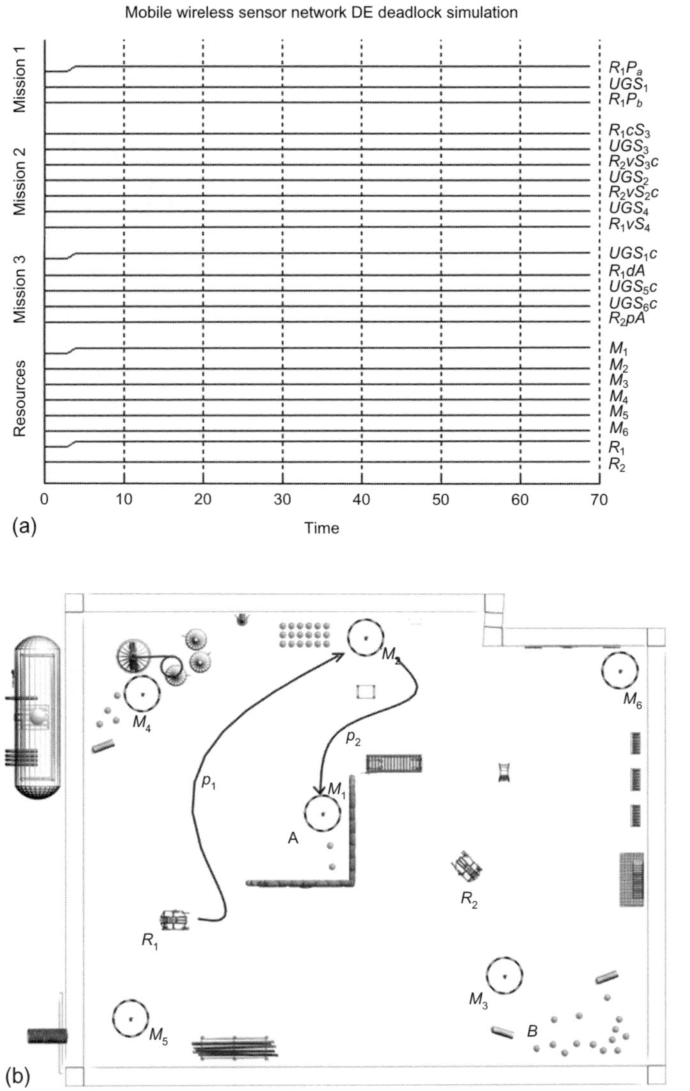


*Figure 6.10 Missions 1 and 3 deadlock: (a) event time traces – MATLAB simulation and (b) top view of robots' paths*

deadlock, and the DE transition traces are as seen in Figure 6.12. On enabling both routing and deadlock avoidance, and triggering the mission multiple times (ten times) to cause multiple complex deadlocks, we get the event trace as seen in Figure 6.13. This clearly illustrates that the mission is taken to completion smoothly without any deadlocks multiple times. Comparing Figures 6.12 and 6.13, we see, as expected, that with multiple choices of routing for a particular task, the overall throughput of the entire mission is greatly improved.

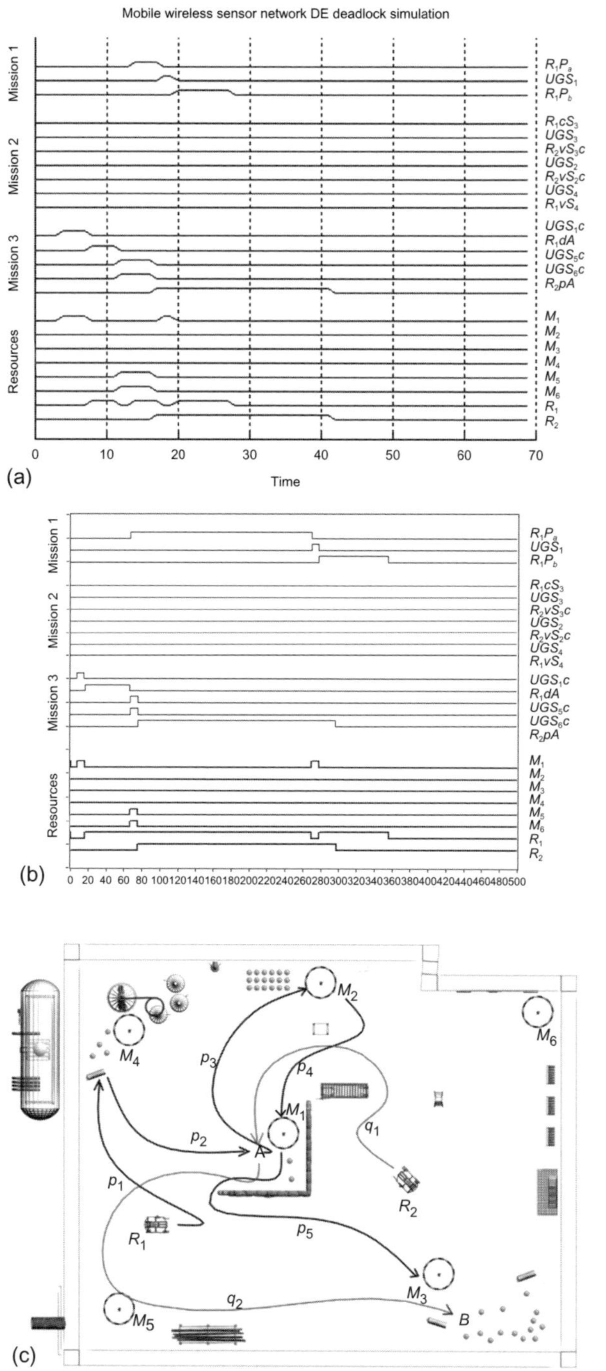


*Figure 6.11* *Missions 1 and 3 deadlock avoidance: (a and b) event time traces – MATLAB simulation and LabVIEW implementation and (c) top view of robots' paths (darker paths are $R_1$ paths; lighter paths are $R_2$ paths)*

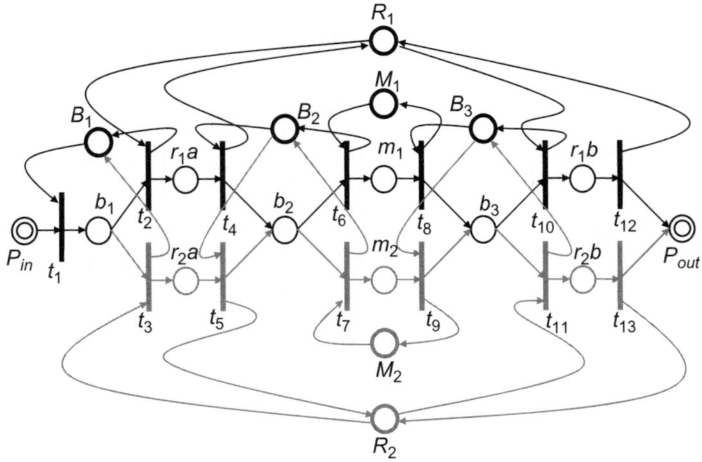


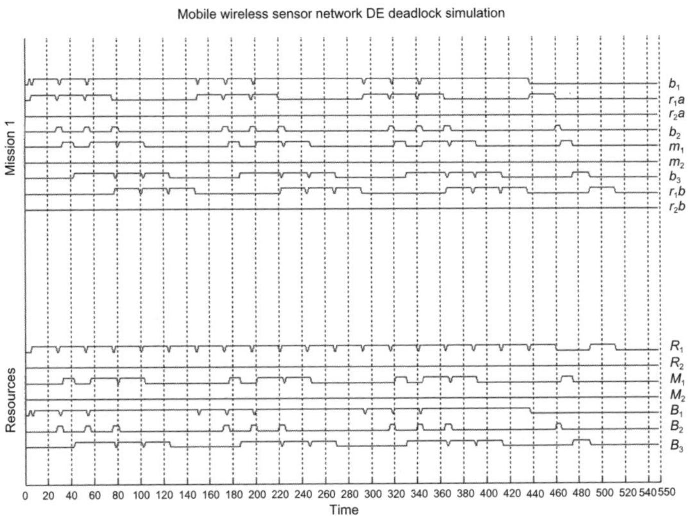


*Figure 6.12 Deadlock avoidance simulation with all routing disabled*

## 6.5 Summary

This chapter has extended the preliminary analysis of deadlock avoidance polices for shared resources in heterogeneous mobile sensor networks to more complicated scenarios. Through experimental implementation on an actual mobile sensor network test bed, we have shown the feasibility and effectiveness of the proposed deadlock-free supervisory control in performing complex and simultaneous sequencing of interconnected tasks. Further, a general deadlock avoidance policy for system with flexible routing, where both shared and routing resources are present, has been mathematically formulated and various simulations performed to validate deadlock-free operation in the presence of multiple routing choices.

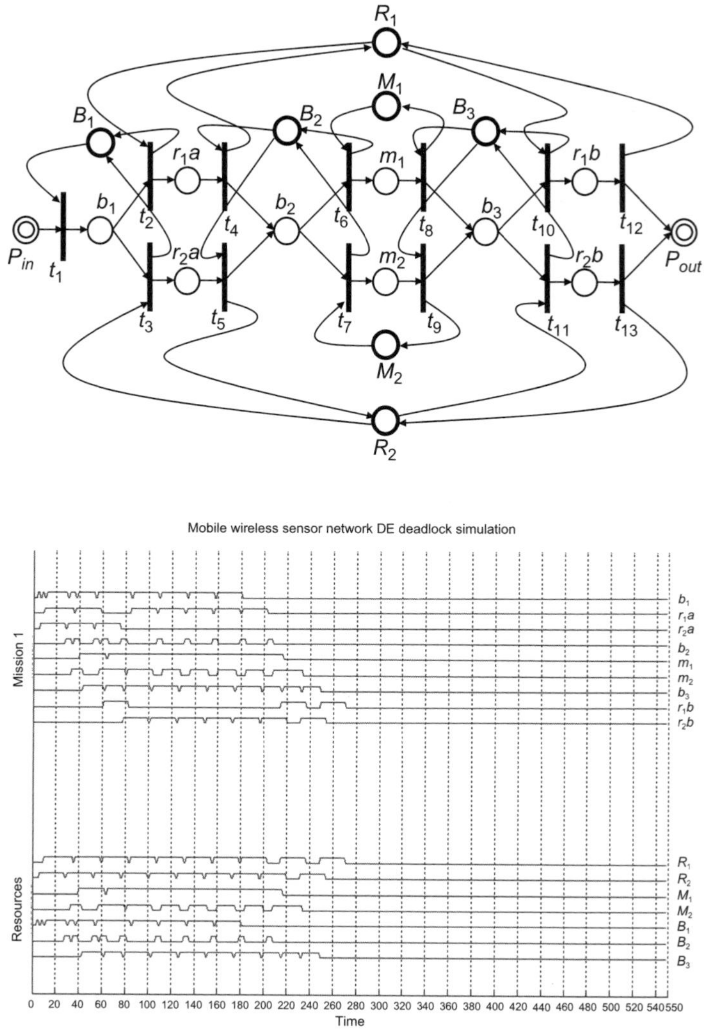


*Figure 6.13 Deadlock avoidance in the presence of routing choices*

# Chapter 7
# Adaptive localization

A range-free approach for adaptive localization of unlocalized sensor nodes employing a mobile robot with GPS is detailed. A mobile robot navigates through the sensor deployment area broadcasting its positional estimate and the uncertainty in its estimate. Distributed computationally inexpensive, discrete-time Kalman filters, implemented on each static sensor node, fuse information obtained over time from the robot to decrease the uncertainty in each node's location estimate. On the other hand, because of dead reckoning and other systematic errors, the robot loses positional accuracy over time. Updates from GPS and from the localized sensor nodes serve in improving the localization uncertainty of the robot. A continuous–discrete extended Kalman filter (CD EKF) running on the mobile robot fuses information from multiple distinct sources (GPS, various sensors nodes) for robot navigation. This two-part procedure achieves simultaneous localization of the sensor nodes and the mobile robot. Also presented in this chapter is an adaptive localization strategy to navigate the mobile robot to the area of least localized sensor nodes. This ensures that the robot manoeuvres to an area where the sensor nodes possess the largest uncertainty in location, so that it can maximize the usefulness of its positional information in best localizing the overall network.

## 7.1 Sensor localization using mobile robot

In this section, we provide an algorithm that runs on each unattended ground sensor (UGS) node that allows it to update its position estimate, and the uncertainty in that estimate, as a mobile robot with known position moves through the network. The algorithm is range-free in that only the communication range need be known, not the range from the node to the mobile robot. It is assumed in this section that the mobile robot's position is exactly known.

### *7.1.1 Scenario*

A deployed wireless sensor network (WSN) composed of static UGS is to be absolutely localized by a mobile robot. The robot broadcasts consist of its own position and its position uncertainty estimates. Broadcasts can only be heard within the robot's communication range. The static sensors, on receiving these broadcasts, combine the new information to update their current location estimate. A simple discrete-time Kalman filter running on each static sensor node serves to fuse information and update its location and uncertainty estimates.

This is a formalized rigorous approach employing Kalman filters for localization, in contrast to bounding boxes [41,42], which are harder to update and keep track of. The developed algorithm is simple and can efficiently be implemented on the sensor nodes with a small computing power. The Kalman filter is simply an optimal recursive data processing algorithm [111] and has been a subject of extensive research and applications, particularly in the area of autonomous navigation.

### *7.1.2 Robot control*

A classical three-wheeled tricycle robot model is employed in all simulations. This configuration uses a controlled steering angle and drive speed to navigate to a desired position as illustrated in Figure 7.1.

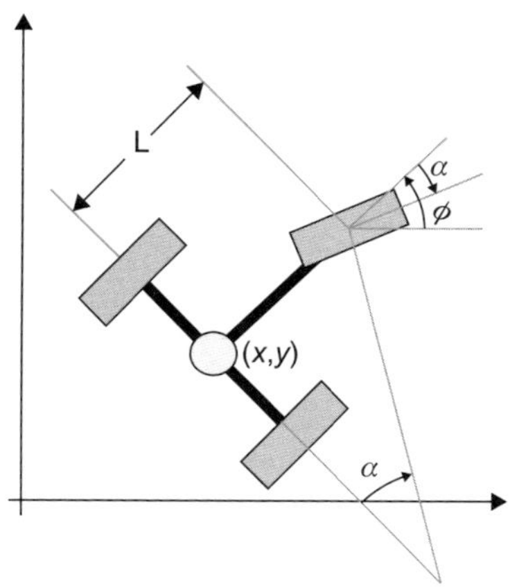


*Figure 7.1 Tricycle robot configuration*

The states and kinematics of the robot are given by

$$X = [x\ y\ \phi\ \alpha]^{\mathrm{T}} \tag{7.1}$$

$$\dot{X} = a(x,t) = \begin{bmatrix} \dot{x} \\ \dot{y} \\ \dot{\phi} \\ \dot{\alpha} \end{bmatrix} = \begin{bmatrix} v_t \cos\alpha \cos\phi \\ v_t \cos\alpha \sin\phi \\ \dfrac{v_t}{L} \sin\alpha \\ \omega_\alpha \end{bmatrix} \tag{7.2}$$

where $(x, y)$ is the position of the robot, $\alpha$ is the steering angle and $\phi$ is the heading angle. The control inputs are the speed $v_t$ and the steering rate $\omega_\alpha$.

A simple proportional-derivative goal-based controller with a temporally varying goal is implemented to navigate the robot along a desired trajectory. For more details, see Reference 112.

This dynamic setup allows more accurate simulations than the simple moving-point model usually assumed in sensor network localization papers.

### 7.1.3 Sensor node Kalman filter

Each static sensor node maintains its own position and uncertainty estimates. The mobile robot broadcasts contain the robot's position estimate and uncertainty estimate. The broadcasts can only be heard within the robot's communication range. A discrete-time Kalman filter running on each sensor node combines this information to optimally update the node's position estimate and its uncertainty. For more details on the derivation of the Kalman filter equations, interested readers are referred to Reference 113.

The Kalman filter is a set of mathematical equations running in a software algorithm that provide an efficient computational means to estimate the state of a process. The state of sensor $i$ at discrete time instant $k$ is

$$x_k^i = [x^i \ y^i]^{\mathrm{T}} \tag{7.3}$$

The sensor state is governed by the following linear stochastic difference equation:

$$x_{k+1}^i = A_k^i x_k^i + B_k^i u_k^i + G_k^i w_k^i \tag{7.4}$$

with measurements given by

$$z_k^i = H_k^i x_k^i + v_k^i \tag{7.5}$$

The random variables $w_k^i$ and $v_k^i$ represent process and measurement noises given by

$$x_0^i = (\overline{x}_0^i, P_{x_0}^i), \qquad w_k^i = (0, Q_k^i), \qquad v_k^i = (0, R_k^i) \tag{7.6}$$

where $(\overline{x}_0^i, p_{x_0}^i)$ denotes a Gaussian noise process with mean $\overline{x}_0^i$ and covariance $p_{x_0}^i$, and $(0, Q_k^i)$ $(0, R_k^i)$ similarly denote Gaussian processes with mean O and covariance $Q_k^i$, $R_k^i$ respectively.

For stationary nodes, the system matrices are given by

$$A_k^i = \begin{bmatrix} 1 & 0 \\ 0 & 1 \end{bmatrix}, \qquad B_k^i = \begin{bmatrix} 0 & 0 \\ 0 & 0 \end{bmatrix}, \qquad G_k^i = \begin{bmatrix} 1 & 0 \\ 0 & 1 \end{bmatrix}, \qquad H_k^i = \begin{bmatrix} 1 & 0 \\ 0 & 1 \end{bmatrix} \tag{7.7}$$

The *a priori* position estimates prior to measurement updates at time $k+1$ are given by the time update equations, which give the effects of time on sensor localization:

$$P_{k+1}^{i\,-} = P_k^i + Q_k^i \tag{7.8}$$

$$\hat{x}_{k+1}^{i\,-} = \hat{x}_k^i \tag{7.9}$$

In these equations, $\hat{x}_k^i$ represents the position estimate of node $i$ at time $k$, while the covariance matrix $P_k^i$ gives the corresponding uncertainty in the position estimate.

The *a posteriori* estimates given a position measurement $z_k$ are given by the measurement update equations, which give the effect of the robot broadcast on sensor localization:

$$P_{k+1}^i = [P_{k+1}^{i\;-1} + H_{k+1}^{i\;\mathrm{T}} R_{k+1}^{-1} H_{k+1}^i]^{-1} \tag{7.10}$$

$$\hat{x}_{k+1}^i = \hat{x}_{k+1}^{i\;-} + P_{k+1}^i H_{k+1}^{i\;\mathrm{T}} R_{k+1}^{-1} (z_{k+1}^i - H_{k+1}^i \hat{x}_{k+1}^{i\;-}) \tag{7.11}$$

The covariance matrices $Q_k^i$ and $R_k$ are design parameters chosen by the engineer. With a zero $Q_k^i$, the uncertainty in location of the sensor $i$ remains constant with time. With an extremely small $Q_k^i$, the localization uncertainty slowly drifts with time. This means that the current measurements from the mobile robot are given more weight than the current node position estimate, which avoids the node's becoming too certain of a position that may be incorrect.

When the robot is in range and the sensor hears the broadcast position of the robot, the measurement update equation is used to combine the new information to improve sensor node position and uncertainty estimates. In this section, the robot is assumed to be perfectly localized. Thus, when a sensor hears a broadcast, it could only be within the communication range of the robot whose position is broadcast. The measurement uncertainty matrix $R_k$ reflects this, and is chosen as

$$R_k = \sigma^{Bot}, \quad \sigma^{Bot} = \begin{bmatrix} \sigma_x^{Bot} & 0 \\ 0 & \sigma_y^{Bot} \end{bmatrix} \tag{7.12}$$

$$\sigma_x^{Bot} = \frac{Range_x^{Bot}}{\sigma_{const}}, \qquad \sigma_y^{Bot} = \frac{Range_y^{Bot}}{\sigma_{const}} \tag{7.13}$$

where $\sigma^{Bot}$ is the uncertainty introduced due to $Range^{Bot}$, the communication range of the robot. We assume the design parameter $\sigma_{const} = 3$ to include 70 per cent of the communication range, $Range^{Bot}$, of the robot. (Gaussian uncertainties are assumed.) Through this selection of $R_k$, the Kalman filter automatically takes care of the range of the robot within which it hears broadcasts.

Algorithm 1 shows the position update algorithm that runs on each node, which is very simple and easy to implement. It consists of four equations, two for the time update and two for the measurement update. This algorithm automatically provides uncertainty estimates through the computation of the error covariance $P_k^i$, which is equivalent to the bounding box information provided by the algorithm in Reference 41.

**Algorithm 1: Static sensor node localization algorithm**

1. *At each discrete time instant,*
2. ***if** robot broadcast received by sensor*
3. ***then***
4. *Update sensor state and uncertainty estimates using KF measurement equations (7.10), (7.11).*
5. ***else***
6. *Propagate estimates using time update equations (7.8), (7.9).*
7. ***end if***

### *7.1.4 Simulation results*

Extensive simulations have been performed to verify the effectiveness of the proposed algorithm. We also studied the effects of initial sweep paths and the robot broadcast interval on sensor localization. The mobile robot is navigated along the desired sweep path, and periodic location information is broadcast. On receiving a broadcast, sensors update their location and uncertainty estimates. This is a range-free procedure that relies on the limited communication range of the robot, and as such, the sensor locations are updated based on the position of the robot. That is, the updated sensor position estimate is a weighted combination of its current location estimate and the current location of the robot. Thus, sensors hearing only one broadcast will have an estimated location that is projected onto the path of the robot.

Figure 7.2 shows the initial sinusoidal sweep path and the position and range of the broadcast with a broadcast interval of five discrete time instants. The desired path is the setpoint trajectory input to the robot controller, and the simulated path is the path along which the controller drives the robot. These paths differ due to the non-holonomic kinematics of a tricycle configuration. Robot broadcasts consist of locations from the simulated path in this scenario. The '$\times$' symbols represent the actual positions of the static sensors that are to be localized. The sensor nodes are randomly deployed in an area of $100 \times 100$ m and do not initially know their actual positions. All the nodes have initial position estimates being the centroid of the deployment area, and an initial uncertainty of infinity, corresponding to complete lack of knowledge of their positions. The sensors have a communication range of 20 m (corresponding to a 20-m range of a MICA mote), and the robot has a communication range of 15 m.

Figure 7.3 illustrates the localized sensors after the robot has made its sweep through the network. The '•' symbols represent the final position estimates of the nodes. The uncertainty of the sensors in their position estimates has been depicted as circles and ellipses representing $3\sigma$ of the uncertainty distribution, assuming Gaussian uncertainties. Note that the sensors always outside the communication range of the mobile robot do not become localized. The sensors that receive more

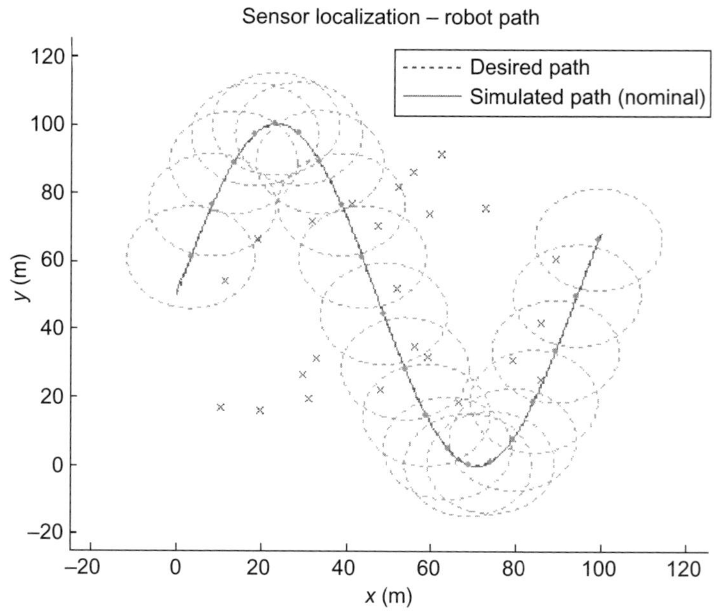


*Figure 7.2 Initial sinusoidal sweep path with broadcast locations and range of broadcast*

than one broadcast from the mobile robot end up being better localized, since each position update reduces the position uncertainty.

The effectiveness of the algorithm is demonstrated by the fact that in every case, the actual location (marked by '×') is within the uncertainty bound of the estimated position (marked by '•' symbols).

The localization error of the sensors, computed as the Euclidean distance between true and estimated positions, is depicted in the vertical axis of Figure 7.4. Sensors near the path of the mobile robot that have received multiple broadcasts have smaller errors.

The same simulation was rerun with different mobile robot broadcast intervals, and the effect of broadcast interval on the average localization error of the network is depicted in Figure 7.5. Generally, as broadcast interval decreases, the average error decreases.

## 7.2 Simultaneous mobile robot and sensor localization

In this section, we consider the realistic case where the mobile robot's position is not exactly known. We provide an algorithm that runs on the mobile robot that fuses position information from GPS, when it is available, and from the already localized sensor nodes. This allows the robot to update its position estimate as well as the uncertainty estimate. When this algorithm is run simultaneously with the

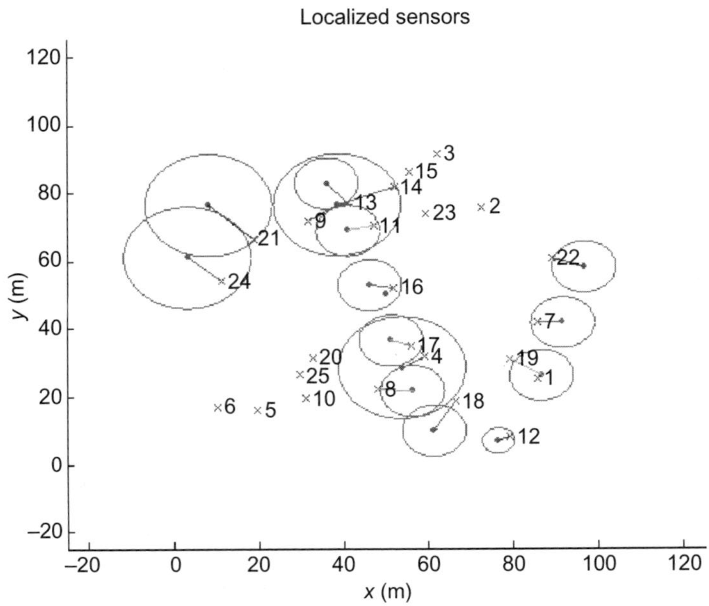


*Figure 7.3* *Localized sensors, real positions (denoted by '×') and estimated positions (denoted by '•') are illustrated after initial mobile robot sweep of the deployment area. Uncertainty ellipses have been illustrated to depict the uncertainty of the sensor in its position estimate*

algorithm of the previous section running on each sensor node, the result is simultaneous mobile robot and sensor localization. A procedure is given to avoid detrimental recursive feedback between the two algorithms.

## *7.2.1 Mobile robot localization*

When localizing the sensor nodes in the previous section, the robot was assumed to know its position exactly at all instants of time. However, as the robot navigates by dead reckoning, or due to steering inaccuracies, its localization increasingly deteriorates as time passes. Location updates from the GPS, when they occur, and from stationary sensor nodes that have already been localized can be used to improve the localization estimate of the robot.

Some sensor nodes are localized more finely due to more numerous updates previously received by them from the mobile robot. These sensors can be employed to localize the robot when its position information deteriorates. This is accomplished by having each sensor node make a transmission that contains the node's position estimate and uncertainty. This is received by the robot when it is in range. The sensors transmit at fixed intervals, with each sensor having a different random interval. This ensures that the updates between mobile robot and sensor nodes are staggered in time and that no recursive feedback occurs. A recursive feedback practically feeds back to the robot its own positional estimate and a slightly larger

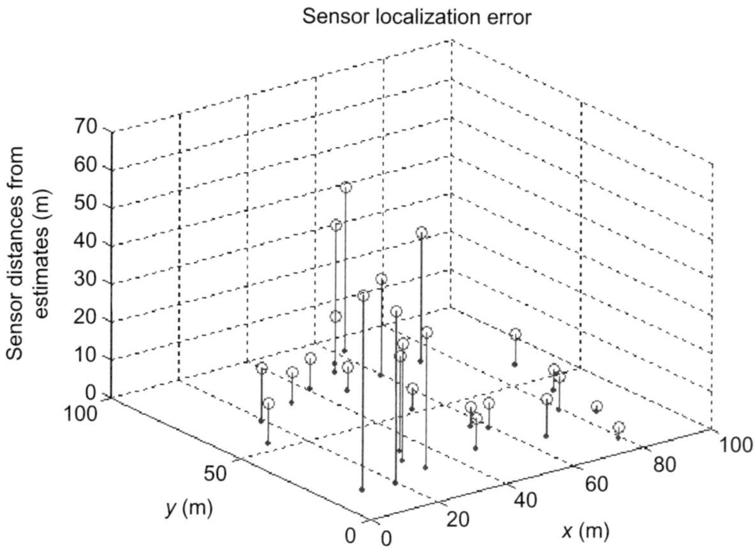


*Figure 7.4 Localization error, computed as the Euclidean distance between real and estimated positions, of sensors after initial sweep of the deployment area*

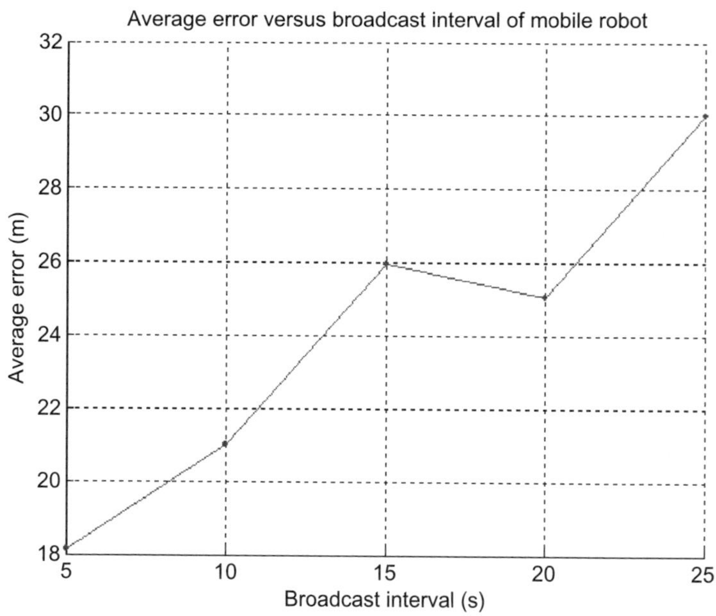


*Figure 7.5 Effect of broadcast interval on average localization error*

uncertainty estimate than its own. This recursive measurement update does not change the positional estimate, but however wrongly reduces the robot's uncertainty estimate. A similar case arises for the sensor if it immediately receives yet another update from the robot.

A CD EKF running on the mobile robot is used to simulate the robot and update the states using measurements from the GPS system and the better localized UGS. Extended Kalman filters (EKFs) have been used for local and infrequent global senor data fusion [114], for mobile robot localization [115] and in navigation of autonomous vehicles [116]. For information about the EKF, see Reference 113.

The continuous-time system model of the robot is given by (7.2) as

$$\dot{X} = a(X, u, t) + G(t)w \tag{7.14}$$

The sampled discrete-time measurement model of the robot is given by

$$\begin{aligned} Z_k^{gps} &= h^{gps}[X(t_k), k] + v_k^{gps} \\ Z_k^{ugs} &= h^{ugs}[X(t_k), k] + v_k^{ugs} \end{aligned} \tag{7.15}$$

where

$$X(0) = (\overline{X}_0, P_0), \quad w(t) = (0, Q), \quad v_k^{gps} = (0, R^{gps}), \quad v_k^{ugs} = (0, R^{ugs}) \tag{7.16}$$

$$a(X, t) = \begin{bmatrix} \dot{x} \\ \dot{y} \\ \dot{\phi} \\ \dot{\alpha} \end{bmatrix} = \begin{bmatrix} v_t \cos\alpha \cos\phi \\ v_t \cos\alpha \sin\phi \\ \dfrac{v_t}{L} \sin\alpha \\ \omega_\alpha \end{bmatrix}, \qquad G(t) = \begin{bmatrix} 1 & 0 & 0 & 0 \\ 0 & 1 & 0 & 0 \\ 0 & 0 & 0 & 0 \\ 0 & 0 & 0 & 0 \end{bmatrix} \tag{7.17}$$

$$h^{gps}[X(t_k), k] = \begin{bmatrix} x \\ y \end{bmatrix}, \qquad h^{ugs}[X(t_k), k] = \begin{bmatrix} x \\ y \end{bmatrix} \tag{7.18}$$

In the EKF, the effect of time on the robot states is given by the time update equation

$$\dot{\hat{X}} = a(\hat{X}, u, t) \qquad \dot{P} = A(\hat{X}, t)P + PA^{\mathrm{T}}(\hat{X}, t) + GQG^{\mathrm{T}} \tag{7.19}$$

In Reference 117, the deleterious effects of time passing are shown in terms of increasing position uncertainty and decreasing belief. These effects are formally captured in a rigorous manner by the time-update equation (7.19), which shows how uncertainty increases due to dead reckoning and steering uncertainties.

The effects of the GPS navigation updates, when they are received, are given by the measurement update equation

$$\begin{aligned} K_k &= P^-(t_k)H^{gps\mathrm{T}}(\hat{X}_k^-)[H^{gps}(\hat{X}_k^-)P^-(t_k)H^{gps\mathrm{T}}(\hat{X}_k^-) + R^{gps}]^{-1} \\ P(t_k) &= [I - K_k H^{gps}(\hat{X}_k^-)]P^-(t_k) \\ \hat{X}_k &= \hat{X}_k^- + K_k[Z_k^{gps} - h^{gps}(\hat{X}_k^-, k)] \end{aligned} \tag{7.20}$$

The effects of the updates based on localized sensor nodes, when they are received, are given by the UGS measurement update equation

$$\begin{aligned} K_k &= P^-(t_k)H^{ugs\mathrm{T}}(\hat{X}_k^-)[H^{ugs}(\hat{X}_k^-)P^-(t_k)H^{ugs\mathrm{T}}(\hat{X}_k^-) + R^{ugs}]^{-1} \\ P(t_k) &= [I - K_k H^{ugs}(\hat{X}_k^-)]P^-(t_k) \\ \hat{X}_k &= \hat{X}_k^- + K_k[Z_k^{ugs} - h^{ugs}(\hat{X}_k^-, k)] \end{aligned} \tag{7.21}$$

The measurement uncertainty matrices $R^{gps}$ and $R^{ugs}$ represent the uncertainty in the GPS and the uncertainty in the update offered by UGS $i$, respectively. The uncertainty in the sensor update, $R^{ugs}$, is a combination of the uncertainty of the sensor position and the uncertainty due to the communication range of the sensor. These uncertainties combine in quadrature as

$$\begin{aligned} R_k^{ugs} &= [P^{i2} + \sigma^{i2}]^{1/2}, \qquad \sigma^i = \begin{bmatrix} \sigma_x^i & 0 \\ 0 & \sigma_y^i \end{bmatrix} \\ \sigma_x^i &= \frac{Range_x^i}{\sigma_{const}}, \qquad \sigma_y^i = \frac{Range_y^i}{\sigma_{const}} \end{aligned} \tag{7.22}$$

where $\sigma^i$ is the uncertainty introduced due to $Range^i$, the communication range of sensor $i$.

Similarly, the measurement noise covariance of the sensor, (7.12), has to be modified to include the uncertainty in the robot's position. The robot is no longer absolutely localized with zero uncertainty. The uncertainty in robot localization and the uncertainty due to robot communication range combine in quadrature, modifying (7.12) as

$$R_k = [P_{XY}^{Bot^2} + \sigma^{Bot^2}]^{1/2} \tag{7.23}$$

where $P_{XY}^{Bot}$ is the partial error covariance of the robot that affects only the position of the robot and $\sigma^{Bot}$ is as defined earlier.

The Jacobians of the non-linear system, determined from (7.2), are given by the following system matrices:

$$A(X,t)=\frac{\partial a(X,t)}{\partial X}=\begin{bmatrix}0 & 0 & -v_t\cos\alpha\sin\phi & -v_t\sin\alpha\cos\phi\\ 0 & 0 & v_t\cos\alpha\cos\phi & -v_t\sin\alpha\sin\phi\\ 0 & 0 & 0 & \dfrac{v_t}{L}\cos\alpha\\ 0 & 0 & 0 & 0\end{bmatrix}$$

$$H^{gps}(X)=\frac{\partial h^{gps}(X,k)}{\partial X}=\begin{bmatrix}1 & 0 & 0 & 0\\ 0 & 1 & 0 & 0\end{bmatrix} \tag{7.24}$$

$$H^{ugs}(X)=\frac{\partial h^{ugs}(X,k)}{\partial X}=\begin{bmatrix}1 & 0 & 0 & 0\\ 0 & 1 & 0 & 0\end{bmatrix}$$

With these equations in place and programmed as a software algorithm on the mobile robot, and the sensor nodes running the algorithm presented in the previous section, the mobile robot and the static sensors automatically mutually update their estimates with incoming updates. There is no additional decision-making logic to be implemented as in other range-free work discussed earlier. There is no need to compute bounding boxes, as the error covariance matrices are automatically updated as measurements are received.

The algorithm to be implemented on the mobile robot that updates its position estimate and uncertainty based on GPS measurements and on the localized sensor nodes is given as Algorithm 2. This algorithm is efficient to implement since the bulk of it is mathematical equations.

When Algorithm 2 is run on the robot simultaneously along with Algorithm 1 on each sensor node, simultaneous mobile robot and sensor localization occurs.

**Algorithm 2: Mobile robot localization algorithm**

1. *Navigate robot along desired path.*
2. *Broadcast location information at discrete intervals.*
3. ***if** broadcast from GPS received*
4. *Update robot state and uncertainty estimates using measurement equation (7.20).*
5. ***end if***
6. ***if** broadcast from sensor received*
7. *Update robot state and uncertainty estimates using measurement equation (7.21).*
8. ***end if***

### 7.2.2 Simulation results

The simulations described in section 7.1 have been rerun with GPS updates and sensor updates implemented as Algorithm 2 on the mobile robot. Infrequent GPS updates and temporally staggered sensor updates help localize the robot. Figure 7.6 shows the robot's sweep path with GPS and UGS updates disabled. A systematic dead-reckoning error [68] has been injected into the mobile robot to give gradually deteriorating position information. The localization of the robot deteriorates with time as can be seen in the deviation of the robot's estimated path (hyphenated grey line) from the robot's true path (continuous grey line). The desired path is as defined earlier, the simulated path is the estimated path of the robot (where the robot thinks it is) and the actual path is the true path of the robot. Robot broadcasts occur along the true path, but the broadcasted position is a corresponding location on the estimated path of the robot. If not corrected, these wrong broadcasts could incorrectly localize the sensors.

Figure 7.7 illustrates the robot's sweep path that is corrected in time by GPS and UGS updates using Algorithm 2. As is evident, the robot's localization has improved and the positions of where the robot thinks it is (the estimated position) and where the robot actually is (the true position) are much closer, since the estimates are continuously corrected using Algorithm 2 as position information arrives, either from GPS or from sensor node broadcasts.

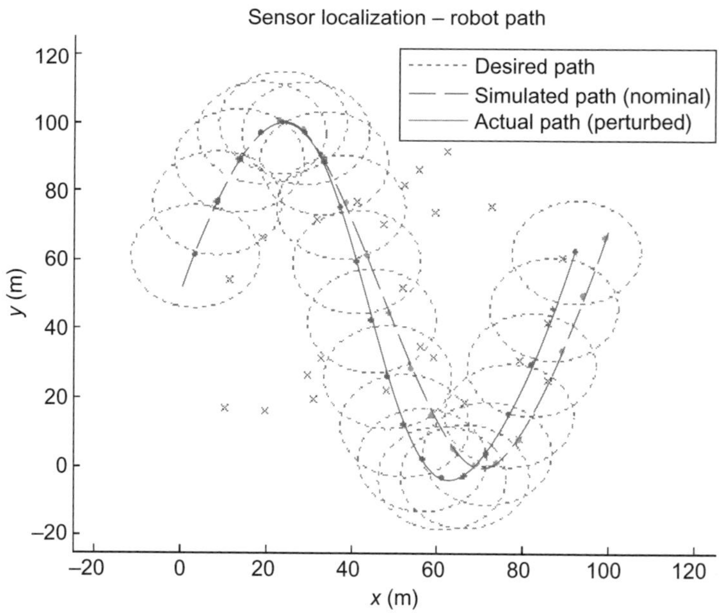


*Figure 7.6 Initial sweep path of the robot with GPS and UGS updates disabled. Robot's localization deteriorates with time as evident in the deviation in the estimated path (hyphenated grey line) and the true path (continuous grey line)*

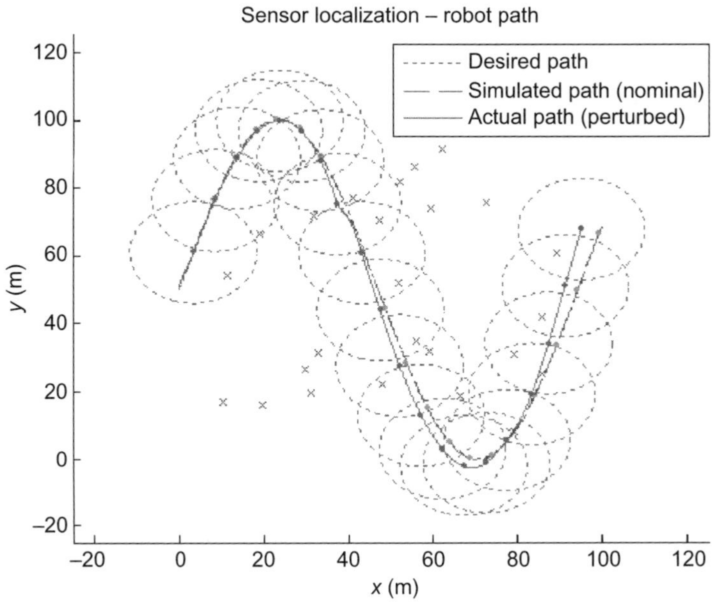


*Figure 7.7 Initial sweep path of the mobile with GPS and UGS updates enabled as Algorithm 2. The robot's localization has improved, and the true estimated positions of the robot along the path are much closer*

Robot broadcasts occur along the true path of the robot and consist of the robot's estimated position (slightly different from the robot's true position where the broadcast occurs) and uncertainty. Sensors within range receive the broadcast and update their positional information based on the robot's estimates.

Figure 7.8 illustrates the localized sensors after the initial sweep. True sensor positions are indicated by '×' symbol and estimated positions by a '•' symbol. Now, some true sensor positions are outside the $3\sigma$ boxes due to the added uncertainty in the robot position, although they are generally close to these boxes. Figure 7.9 depicts the final localization error of each sensor.

Figure 7.10 compares the average localization error of the sensor network obtained by running Algorithms 1 and 2 with UGS and GPS updates off and on. As can be seen, for the case of Algorithm 1 where the robot is assumed to be perfectly localized, the average localization error is less than that for Algorithm 2. And in the case of Algorithm 2, better localization is obtained when the robot receives updates from the UGS and the GPS system.

## 7.3 Simultaneous adaptive localization

A navigation strategy, to be used subsequent to the initial sweep of the deployment area that was presented in the previous sections, is developed here that further minimizes the localization uncertainty of the sensor network in the most efficient

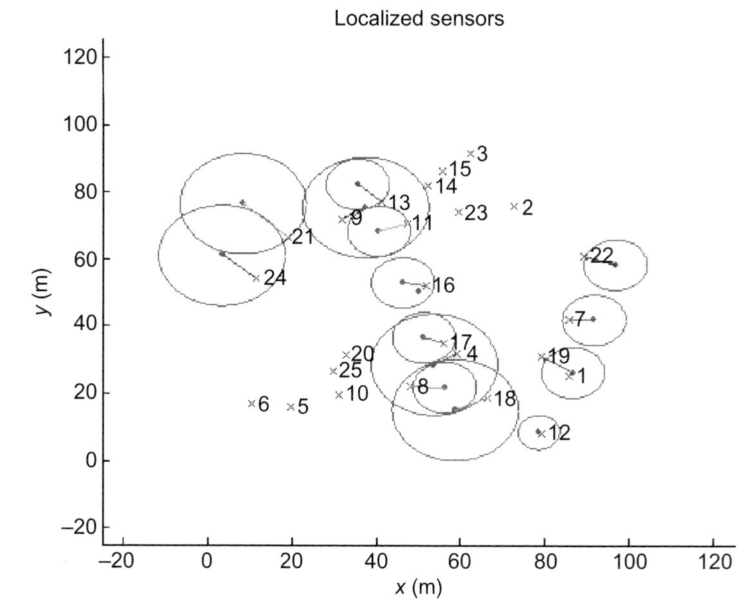


*Figure 7.8 Localized sensors after initial sweep of the deployment area. True sensor positions are indicated by a '×' and estimated positions by a '•'*

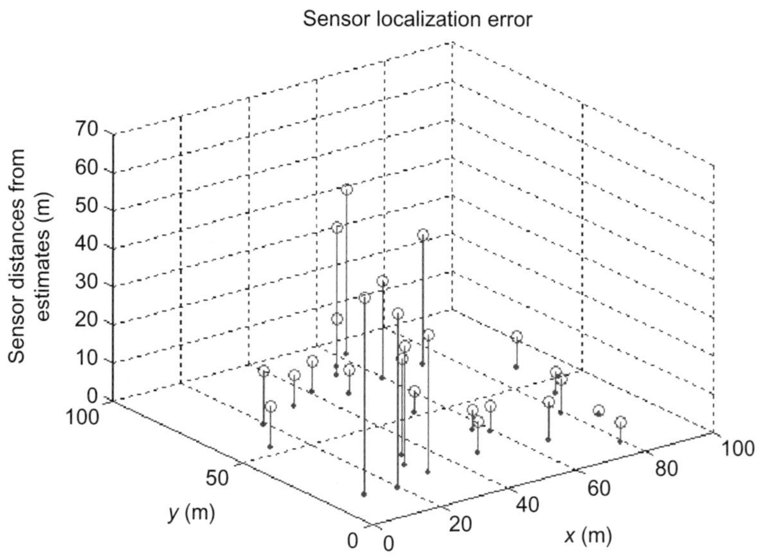


*Figure 7.9 Localization error of sensors computed as the Euclidean distance between true and estimated positions*

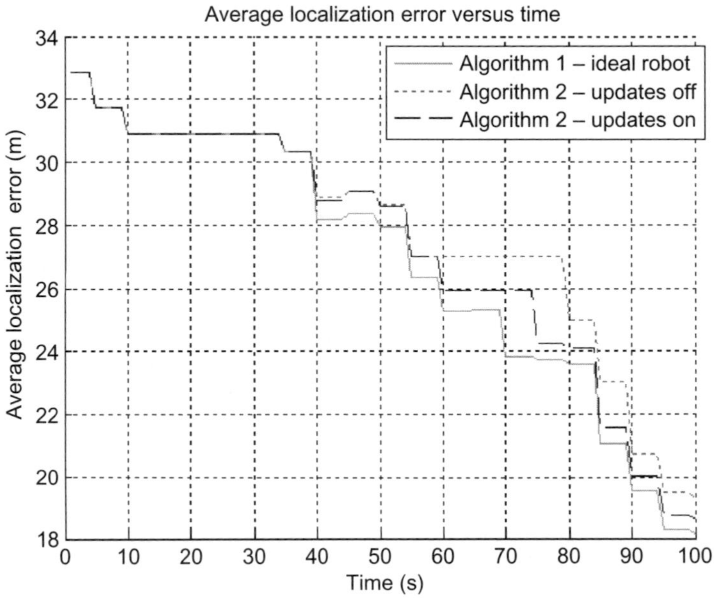


*Figure 7.10 Average localization error over time for Algorithms 1 and 2 with UGS and GPS updates off and on*

manner. An adaptive localization policy is adopted to navigate the mobile robot to an area of least localized sensor nodes. This ensures that the robot manoeuvres to an area with sensor nodes possessing the largest uncertainty in location.

Accurate position of coarsely localized sensors cannot be known (due to inherent coarse localization) so that navigating to these sensors is not possible. The radio connectivity of the network is exploited to address the problem of having the robot navigate to a location that is imprecise. Figure 7.11 depicts the communication connectivity of the network.

A communication protocol is developed wherein the robot broadcasts a navigation request packet, *NAV-REQ*, when the robot wants to find a new location to navigate to. Sensors that receive the *NAV-REQ* packet forward it along the network. Sensors having a large uncertainty scalar, the Frobenius norm [118] of the uncertainty matrix, reply back with a localization request packet, *LOC-REQ*. The *LOC-REQ* packet consists of the uncertainty matrix of the requesting sensor and propagates along the network until it is received by a friendly localized neighbour. Friendly localized neighbouring sensors receiving the *LOC-REQ* packet append it with their position and forward the packet along the sensor network to the robot. Figures 7.12 and 7.13 show the flow of the *NAV-REQ* and *LOC-REQ* packets.

The robot receives packets from multiple non-unique friendly neighbours, each representing a single coarsely localized sensor. The robot needs to choose a friendly neighbour to navigate to. Friendly neighbour arbitration is performed by grouping uncertainties of the same friendly neighbour in quadrature to give its

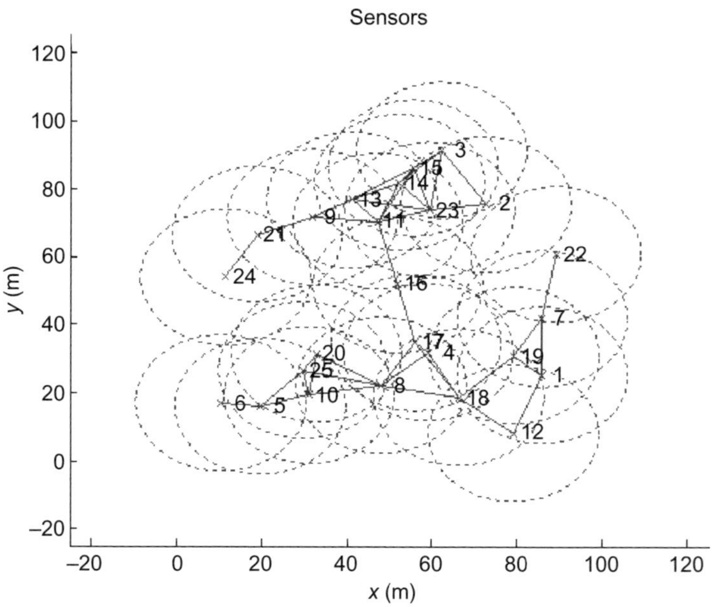


*Figure 7.11 Communication connectivity of the network. Communication routes between sensors and range of communication of each sensor are depicted*

combined uncertainty scalar. The friendly neighbour with the largest combined uncertainty scalar is picked as the location to navigate to. If multiple such neighbours exist, the most localized neighbour is chosen. The trade-off between combined uncertainty and localization certainty of a friendly neighbour even when there is no tie is such that the robot always navigates to reduce the maximum amount of uncertainty with each iteration. This may result in more coarsely localized sensors than one or two finely localized sensors.

Thus, regions with a large density of coarsely localized sensors having a common friendly neighbour are adaptively navigated to. However, due to the inherent imprecise location of the friendly neighbour, the robot actually navigates a circular path around the neighbour's estimated position.

Algorithm 3 summarizes the adaptive localization algorithm.

**Algorithm 3: Adaptive localization algorithm**

1. *Broadcast Navigation request, NAV-REQ, packet.*
2. *Wait to receive Localization request, LOC-REQ, packets.*
3. ***for** all LOC-REQ with the same friendly neighbor*
4. *Combine uncertainty scalars of the requesting sensors.*
5. ***end for***

6. *Pick friendly neighbor with maximum combined uncertainty scalar of the requesting sensors.*
7. ***if** multiple maximas arise*
8. *Among the maxima, pick the most localized friendly neighbor.*
9. ***end if***
10. *Navigate around the picked friendly neighbor executing the simultaneous localization algorithm, Algorithm 1 on the sensors and Algorithm 2 on the mobile robot.*
11. *Repeat Steps 1–10 as required.*

After the initial sinusoidal sweep (see Figures 7.8 and 7.13), both sensors 8 and 14 receive three localization request packets each from sensors 10, 20 and 25, and 3, 15 and 23, respectively. And on combining the uncertainties of the requesting coarsely localized sensors, an equal maximum uncertainty scalar arises for sensors 8 and 14. However, sensor 8 is more localized than sensor 14, and robot navigation occurs around the estimated position of sensor 8 (see Figure 7.14b).

Figure 7.14 illustrates four adaptive localization iterations; robot navigation path is shown in (a–e), and corresponding uncertainty scalars of the sensors at the end of each adaptive localization iteration are illustrated in (f–j). With each adaptive localization iteration, Figure 7.15 shows the reduction of localization error of each sensor, and Figure 7.16 depicts the reduction of the average localization error

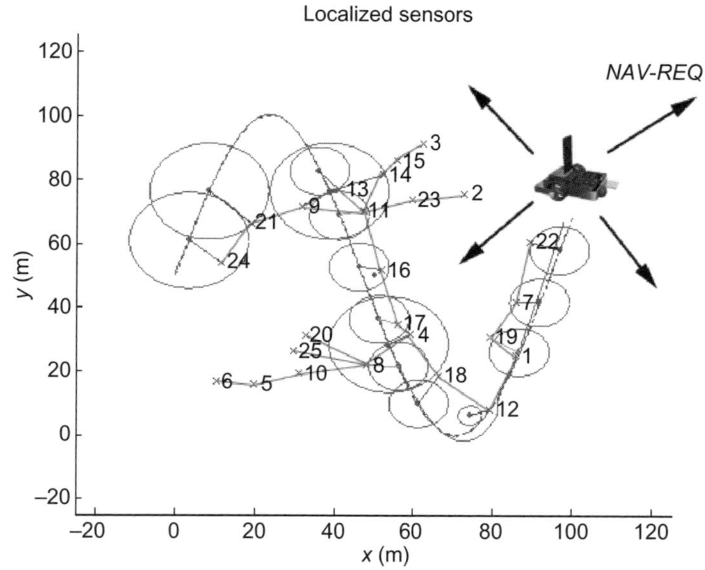


*Figure 7.12 Initiation of the navigation request 'NAV-REQ' packet from the robot*

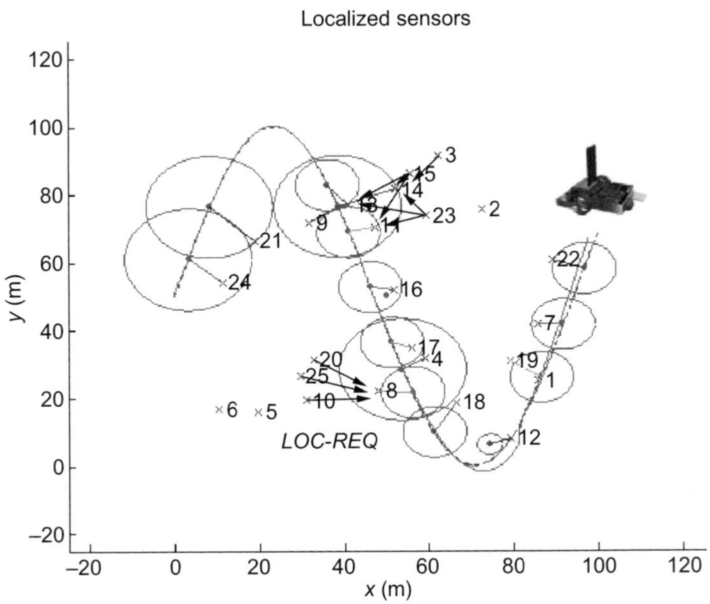


*Figure 7.13 Coarsely localized sensors reply back with a localization request 'LOC-REQ' packet. Black arrows indicate the flow of the 'LOC-REQ' packets to friendly neighbour nodes that append the packet with their position and forward it along the network towards the robot*

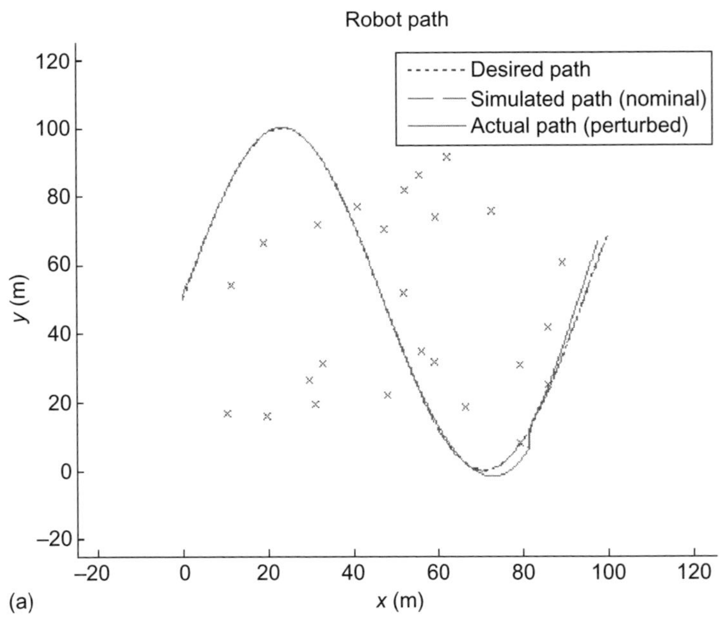


*Figure 7.14 Various adaptive localization steps (a–e) and corresponding uncertainty scalars for the sensors (f–j) after each navigation step*

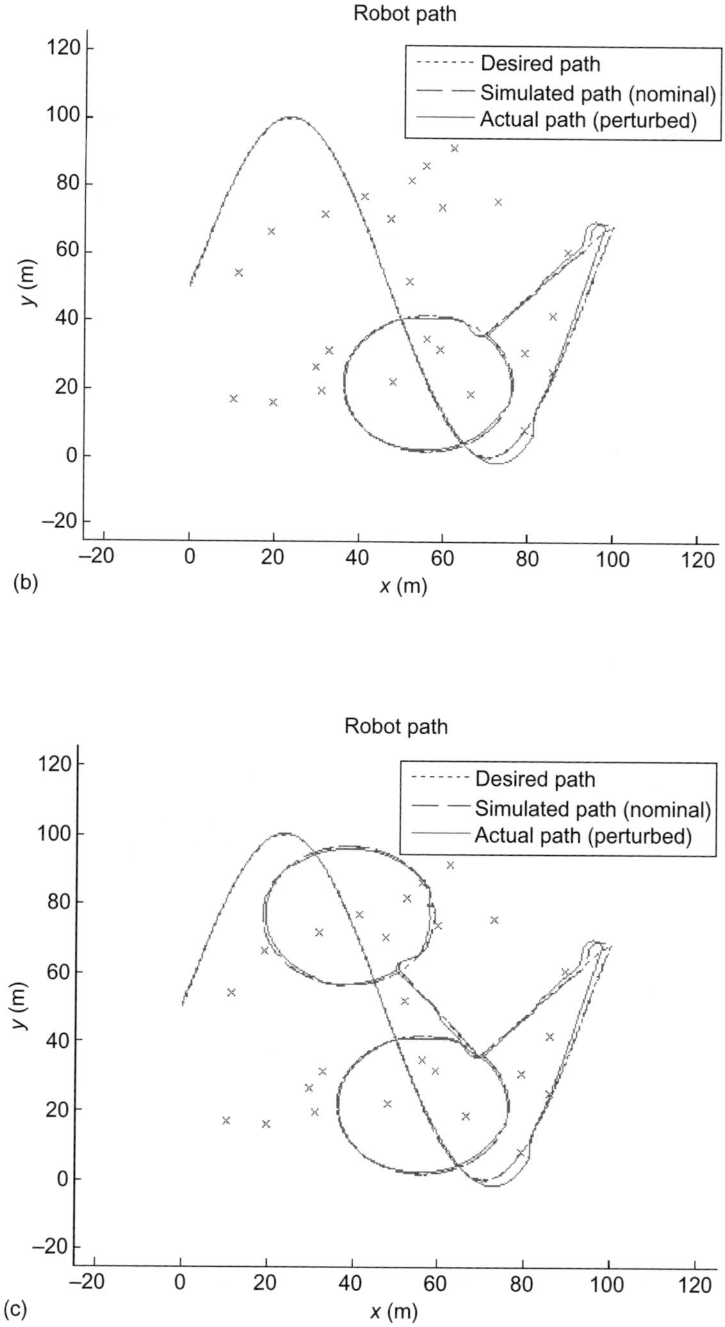


*Figure 7.14 Continued*

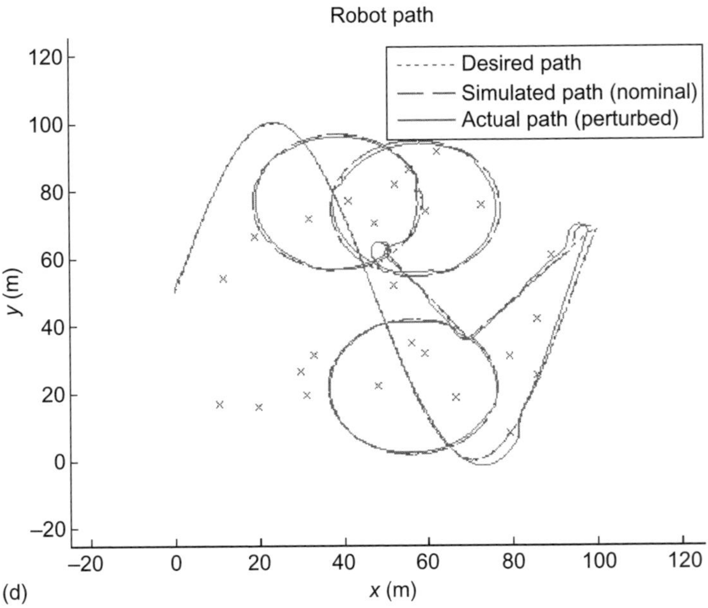


(d)

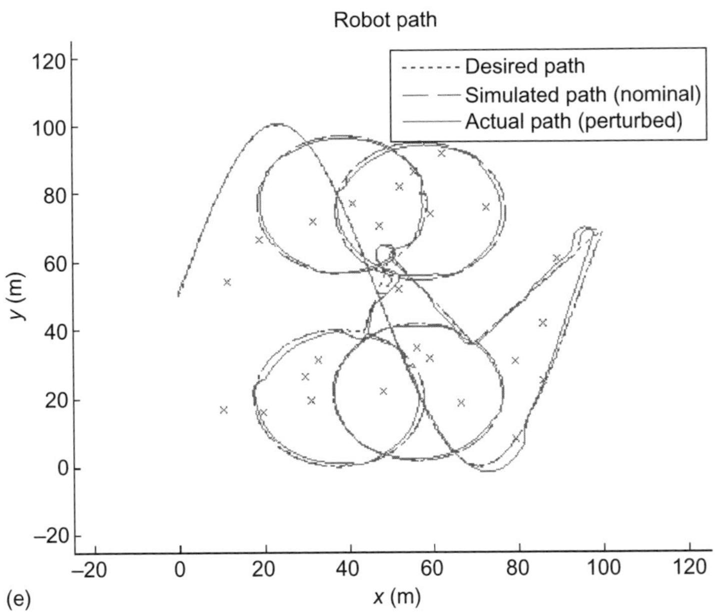


(e)

*Figure 7.14 Continued*

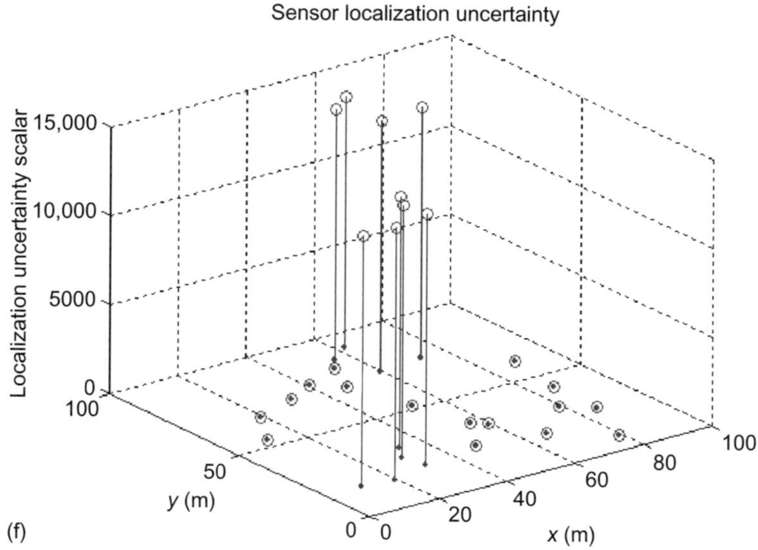


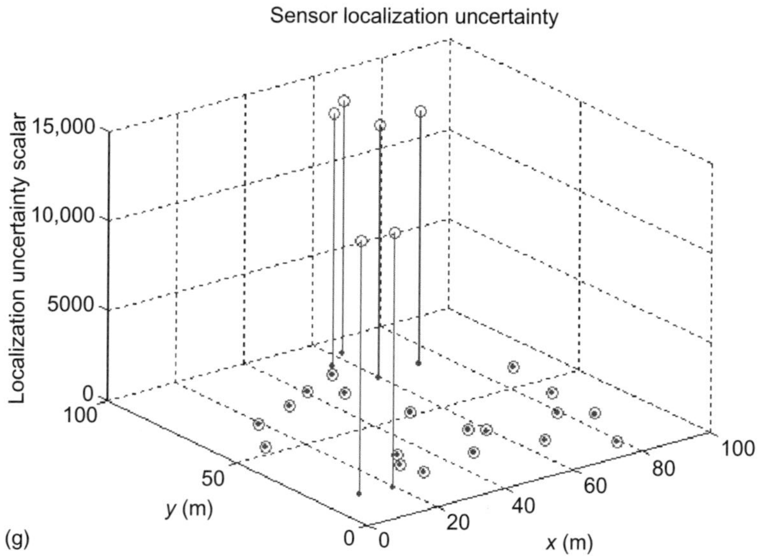


*Figure 7.14 Continued*

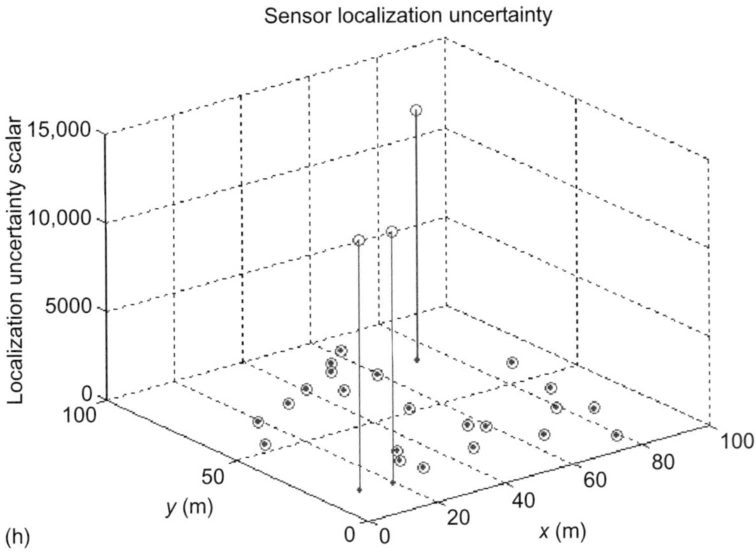

(h)

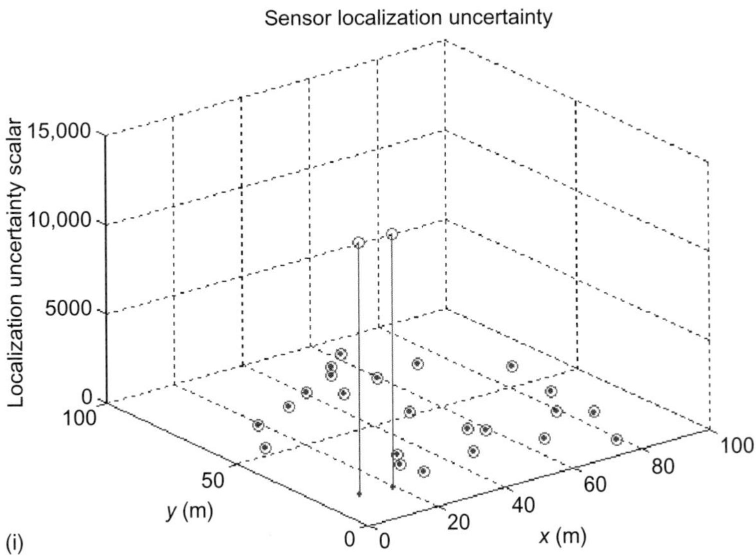

(i)

*Figure 7.14 Continued*

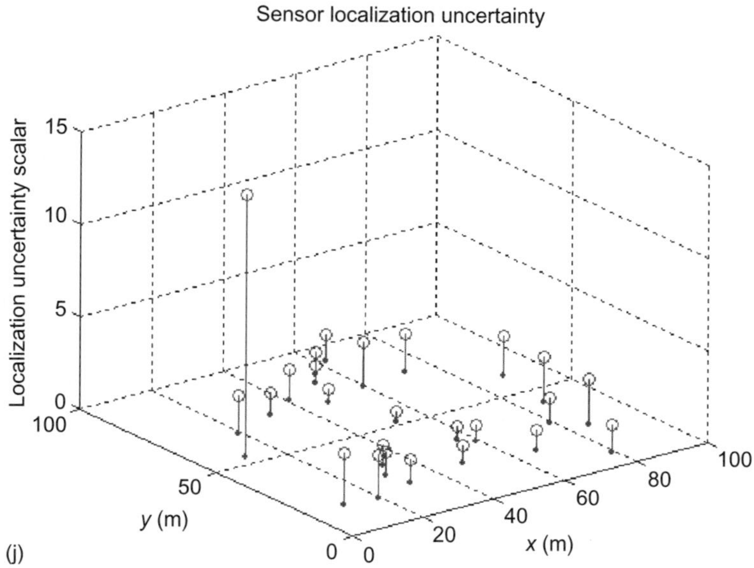


*Figure 7.14 Continued*

of the sensor network. Figure 7.17 illustrates the localized sensors after four iterations of the adaptive localization algorithm. As can be seen, all sensors are localized and uncertainty in localization is fairly small. Figure 7.18 compares the average localization error for the three algorithms (1–3) over time. As can be seen,

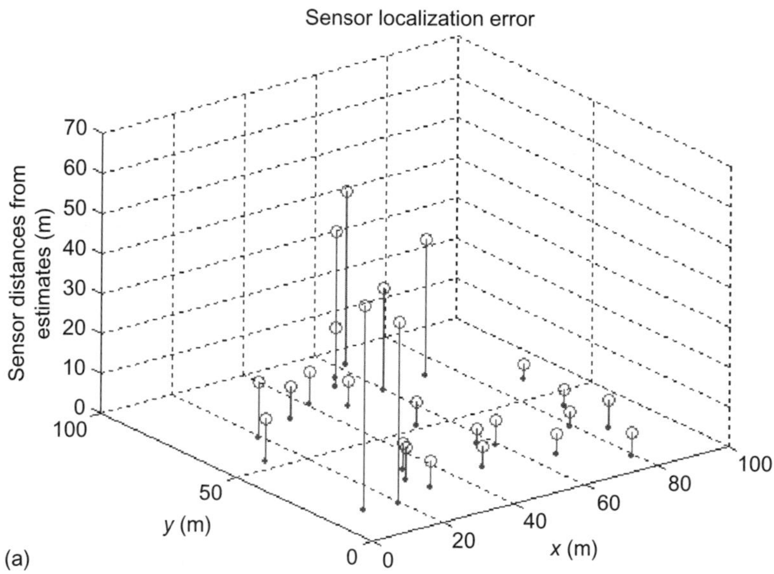


*Figure 7.15 (a–d) Reduction of the average localization error of the sensors with each adaptive localization iteration*

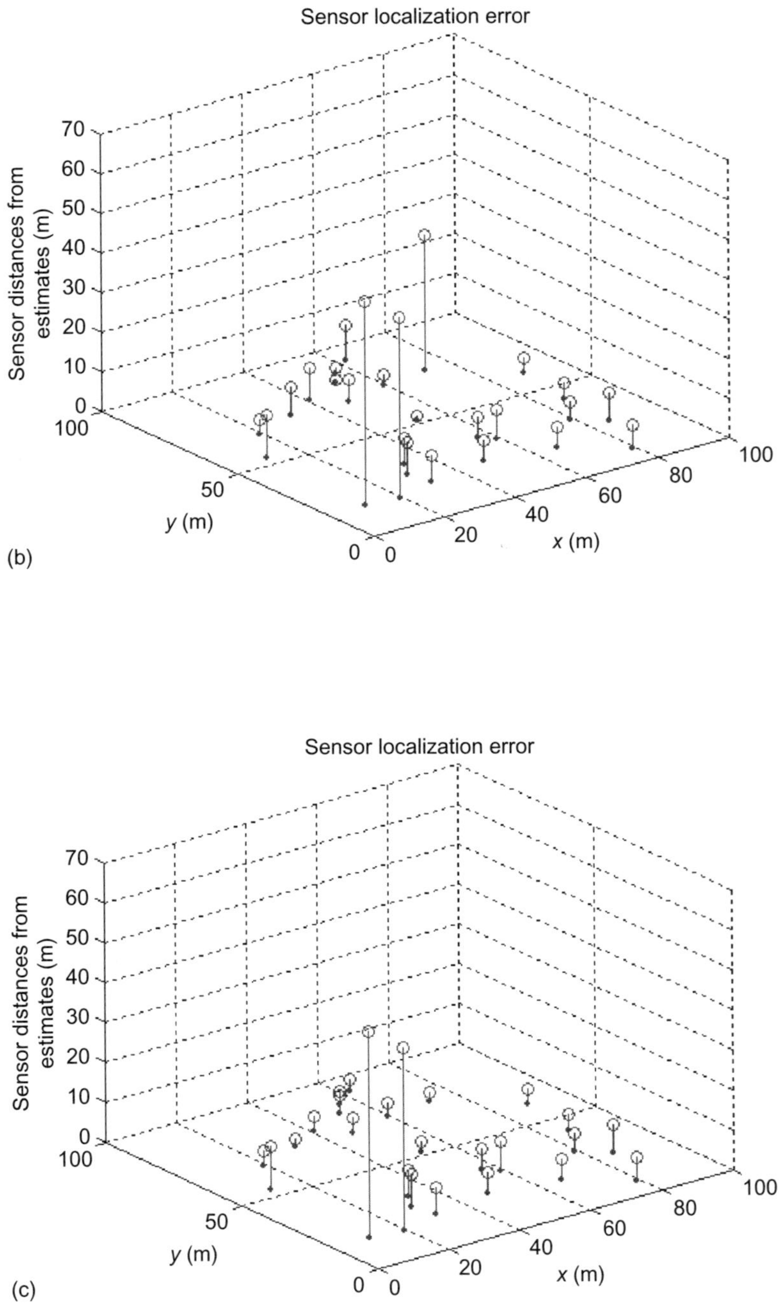


*Figure 7.15 Continued*

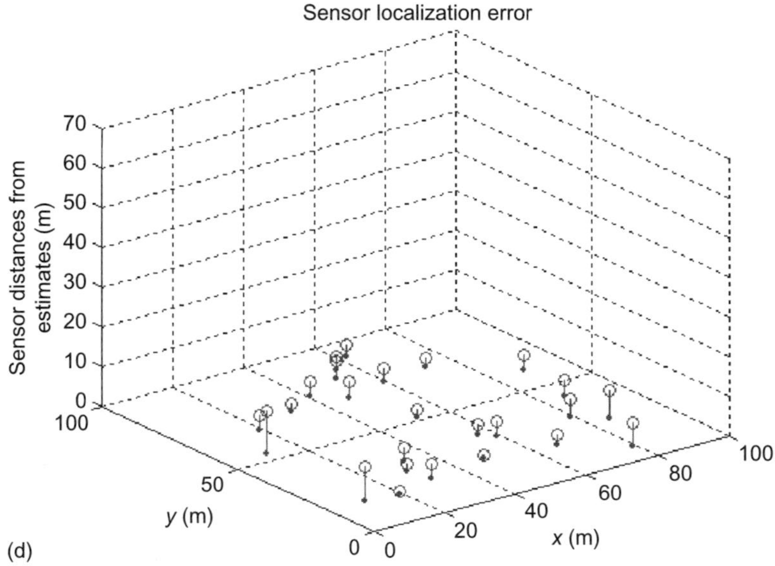


*Figure 7.15 Continued*

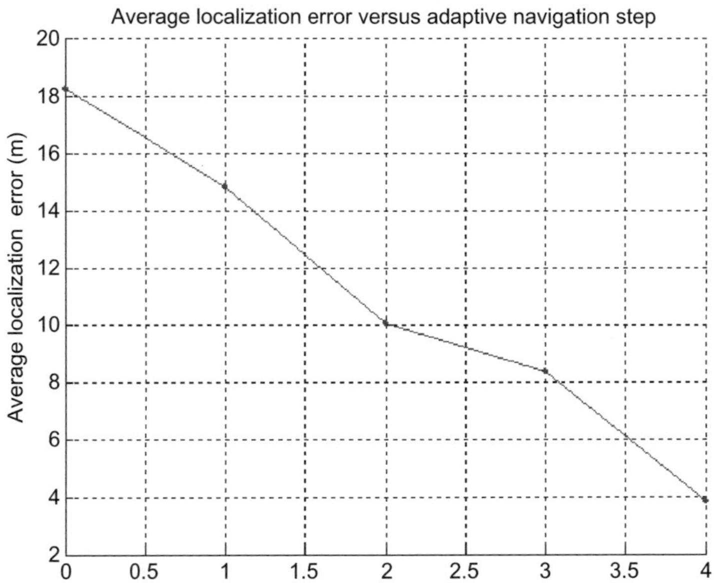


*Figure 7.16 Reduction of average localization error of the sensor network with each adaptive localization iteration*

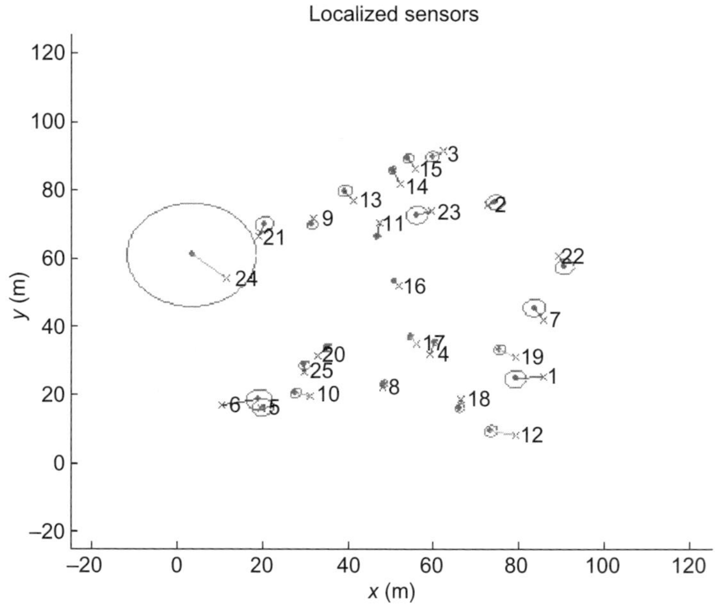


*Figure 7.17 The final position estimates of the localized sensors after four iterations of adaptive localization algorithm*

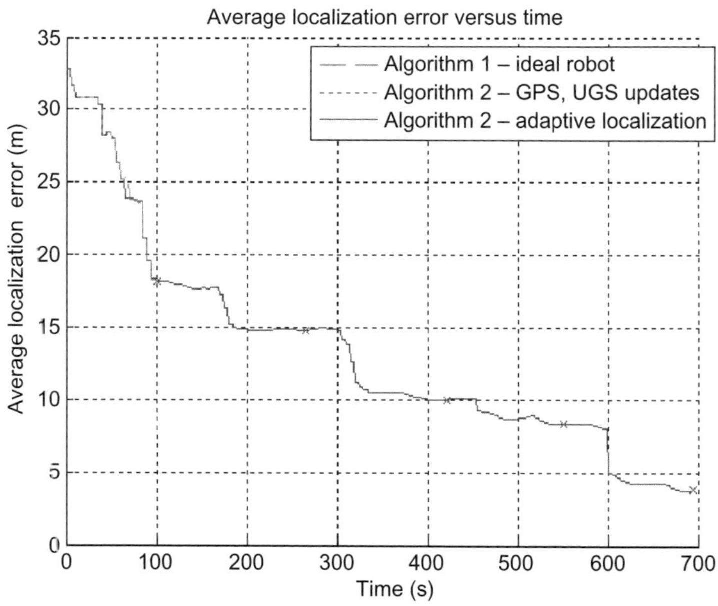


*Figure 7.18 Average localization error of the sensor network with time for Algorithms 1–3. The '×' symbol demarcates each adaptive localization iteration region*

with the adaptive navigation (Algorithm 3), there is a significant reduction in localization error.

At every instant, along with the adaptive localization algorithm, Algorithm 3, the entire simultaneous localization algorithm with updates from the GPS and more localized sensor, Algorithms 1 and 2, are always running. This demonstrates simultaneous adaptive localization of the sensor network.

## 7.4 Extensions

This section discusses how the results of this work are affected by dropping certain ideal assumptions.

### *7.4.1 Effect of uncertainty matrices*

Infrequent accurate GPS updates are assumed to exist that, along with updates from the UGS, help in localizing the robot. Figure 7.19 (see p. C16) shows the effect of the GPS uncertainty matrix, $R^{GPS}$ (7.20). With a highly inaccurate GPS (large $R^{GPS}$), the localization of the sensor network initially decreases as sensors get localized, but then increases again as the robot drifts are wrongly corrected and the sensors get localized to the increasingly incorrect robot broadcasts.

On the other hand, with increasing robot uncertainty matrix, the robot broadcast locations are accurate but with large uncertainty. Figure 7.20 shows the effect of the robot uncertainty matrix through $Q$ (7.19) on localization. As expected, localization error is very slightly affected; however, the uncertainty of the sensors in their estimates increases with $Q$.

### *7.4.2 Effect of radio range and irregularity*

The localization algorithms presented here entirely rely on the sensor being able to hear robot broadcasts. Figure 7.21 (see p. C16) shows the effect on localization by varying the communication range of the robot only (the range of the sensors is fixed at 20 m). An extremely small robot range does not actually improve localization as sensors hear either a small number of robot broadcasts or none at all. With large robot ranges, all sensors are rapidly localized but with coarse localization. A robot communication range that is slightly smaller than the sensor communication range seems to obtain the best localization results.

Radio irregularity is a common and non-negligible phenomenon in WSNs. It results in irregularity in radio range in different directions [119]. A Radio Irregularity Model (RIM) discussed in Reference 120 has been implemented to see the effect of radio irregularity on sensor network localization. Figure 7.22a shows the robot communication range for different values of degree of irregularity (DOI). Figure 7.22b shows the robot broadcasts with the implemented dynamic RIM where the range is asymmetric and it dynamically reorients based on the orientation of the robot.

Algorithms 1–3 have been run ten times for each of the different DOIs, and the mean sensor network localization error is obtained. The number of adaptive

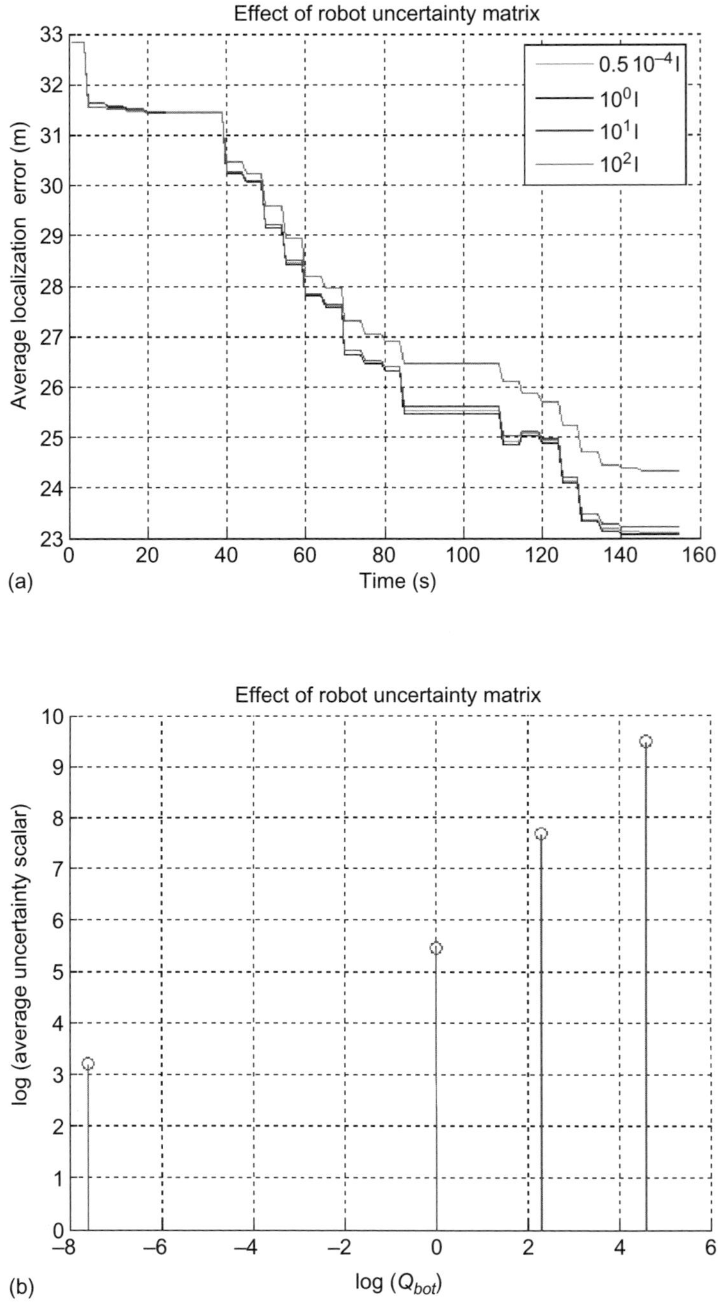


*Figure 7.20 Effect of robot uncertainty matrix on (a) average sensor network localization and (b) mean uncertainty scalar in sensor network localization*

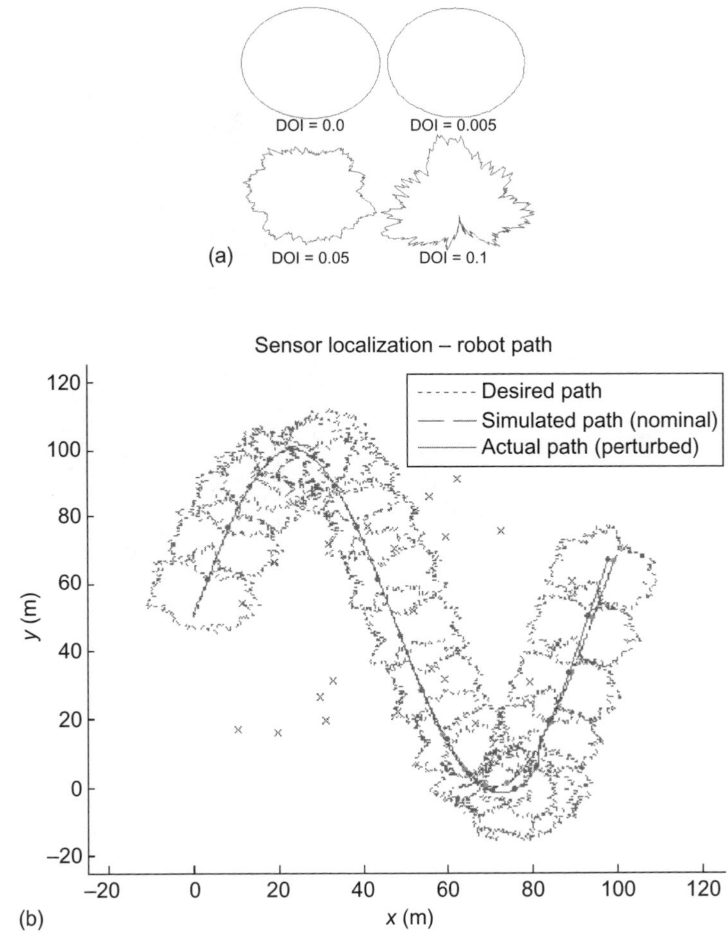


*Figure 7.22* *(a) Radio communication range for different DOIs and (b) robot broadcasts with the dynamic RIM*

localization iterations has been limited to four to provide means of a fair comparison. As seen in Figure 7.23, with increasing radio irregularity, the localization error deteriorates. Certain robot broadcasts may be lost due to the radio irregularity; however, adaptive localization continues to work as long as the network does not break up into trivial clusters of a single node. For an increased DOI, the algorithm will just take more iterations to converge to an equivalent localization error obtained with no radio irregularity.

One aspect that has not so far been addressed is related to the issue of drop packets/link connectivity. In other words, what happens if NAV-REQ and LOC-REQ packets are received intermittently? Research [121] has shown that nodes may be in communication range but yet occasionally drop packets due to environmental factors, interference, congestion and other sources of loss. It is possible to keep

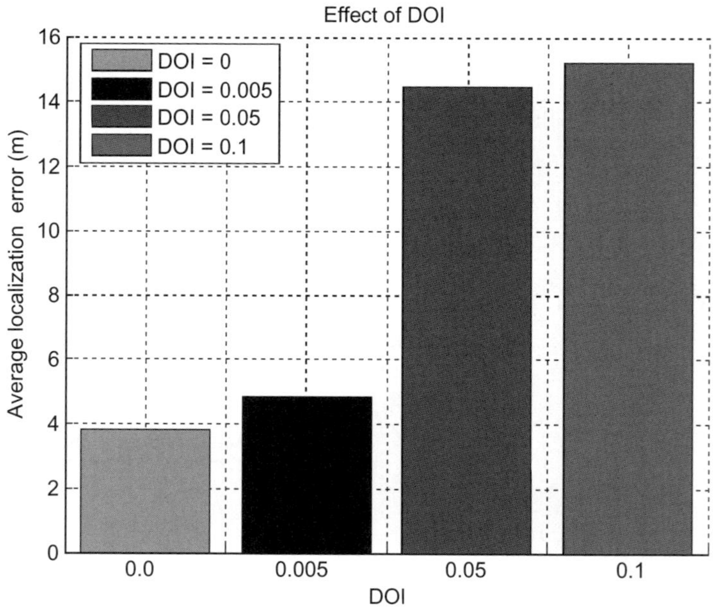


*Figure 7.23 Effect of radio irregularity on sensor network localization*

retransmitting packets and increase the delivery probability, but this also increases the transmission time and energy consumed. Figure 7.24 shows the effect on the adaptive localization algorithms when communication links fail with certain probabilities and packets are dropped. The mean sensor localization error deteriorates as links fail and packets get dropped.

### 7.4.3 Energy considerations

Algorithm 1 on the sensor node, although computationally inexpensive, places the sensor in listen mode for extended periods of time. This is a case of idle listening, where the sensor node actively listens for potential robot broadcasts. Idle listening of the radio has been identified as a major source of energy wastage [122]. A survey of energy-saving mechanisms in sensor networks exists in Reference 123.

In our application, since all communications for our algorithms are discrete in nature, the listening of the nodes can be discretized. For instance, if the knowledge that the robot broadcasts every 5 s is available to every node, then the node can sleep for 4 s and listen for 1 s. This in itself is an energy saving of over 80 per cent. All NAV-REQ and LOC-REQ packet transmissions can by synced to this interval. Periodic correction of clock drifts for various nodes would however be required. In the initial case of a node not knowing when to listen, it can listen 100 per cent of the time until it finds out the discrete broadcast periodic interval. For detailed information on sleep scheduling, refer to References 124 and 125.

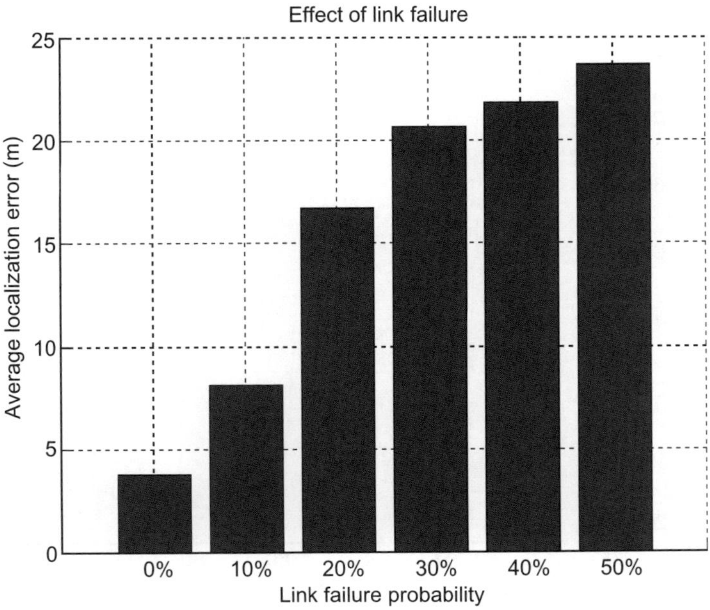


*Figure 7.24 Effect of link failure on sensor network localization*

## *7.4.4 Extensions to simultaneous adaptive localization*

Sensor nodes are localized only by broadcasts from the mobile robot. Other sensor broadcasts could be used for localization, but for even a reasonably fine localization, multiple localized neighbours around the sensor are required. Multiple broadcasts from one particular sensor provide as much information as just one broadcast.

The initial sweep path of the mobile robot over the sensor network may place certain restrictions (such as knowing the extent of the sensor network to be swept). However, this sweep path need not cover the entire area; in fact, the sweep path can be entirely dropped and the adaptive localization algorithm made to completely take over. An initial NAV-REQ would result with an LOC-REQ and will navigate the robot to the initial estimate of the sensor requesting localization (usually the centre of the deployment area, but could be any other location). Figure 7.25 shows four adaptive localization iterations with no initial network sweep. As shown in Figure 7.26, with no initial sweep, the localization error initially is high, but eventually converges to that obtained with an initial sine sweep.

The distributed nature of Algorithms 1–3 also presents the ability of trivially introducing additional mobile robots into the network. Each robot can implement Algorithms 2 and 3 with absolutely no changes, thereby accelerating localization of the sensor network to an even fine granularity.

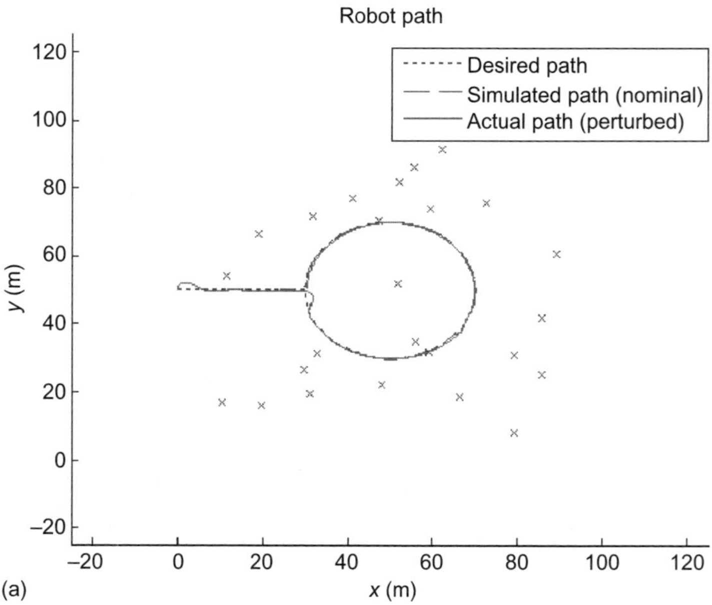


(a)

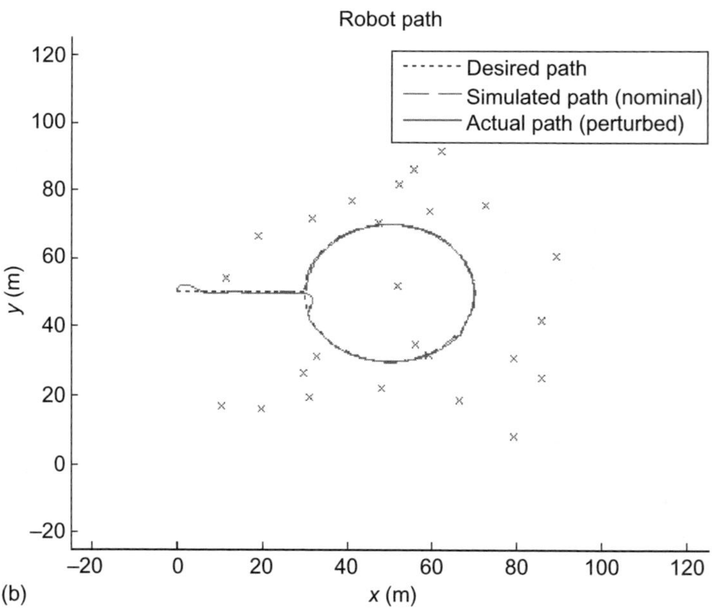


(b)

*Figure 7.25 (a–d) Adaptive localization paths with no initial sweep*

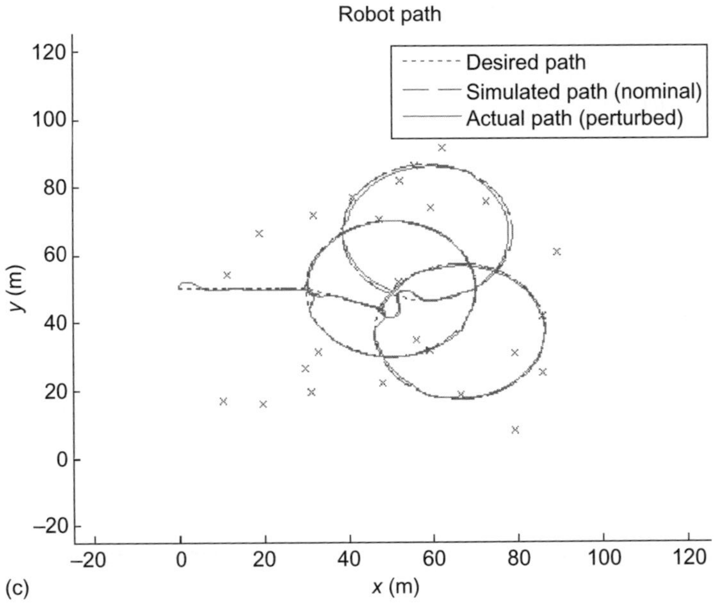


(c)

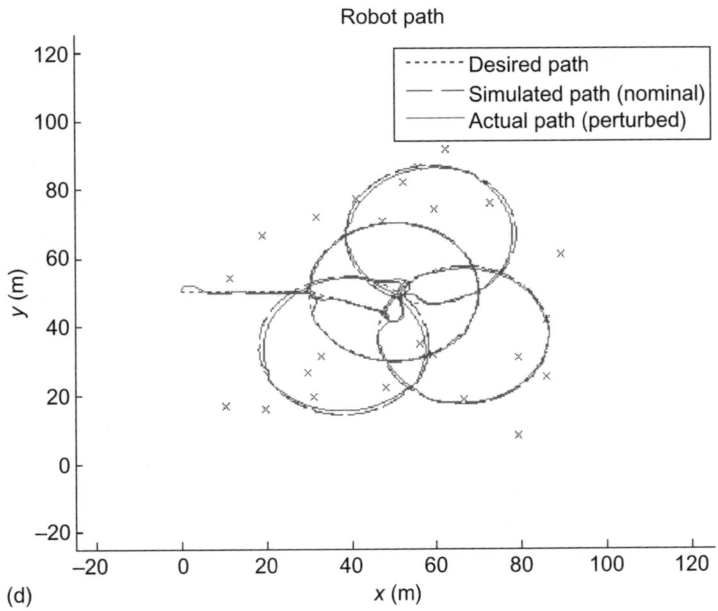


(d)

*Figure 7.25 Continued*

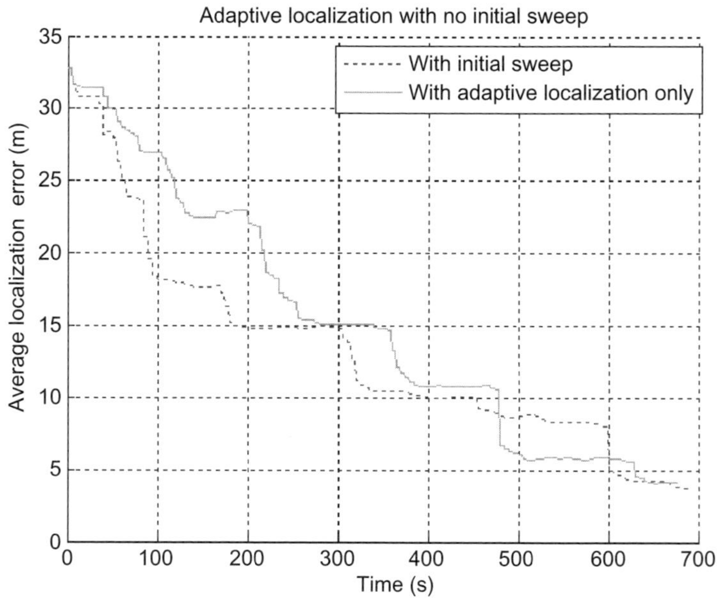


*Figure 7.26 Adaptive localization error with no initial sweep*

## 7.5 Summary

Rigorous mathematical algorithms for adaptive simultaneous localization of the static UGS and the mobile robot have been demonstrated. The first algorithm localizes the static sensors, and the second algorithm localizes the mobile robot. These algorithms together allow simultaneous localization of the static sensor and the mobile robot. A third adaptive localization algorithm ensures that the region of the deployment area with the largest uncertainty is localized with minimal robot movement.

Experimental validations of the proposed algorithms using the Automation & Robotics Research Institute's (ARRI) WSN test bed and the mobile robot developed by the authors in Reference 50 are to be carried out. A communication link estimation model as discussed in Reference 121 may need to be added to the Kalman filter to reliably route NAV-REQ and LOC-REQ packets in an efficient manner in the presence of lossy communication links.

# References


1. C. Huang and Y. Tseng, 'The coverage problem in a wireless sensor network', in *Proceedings of the 2nd ACM International Conference on Wireless Sensor Networks and Applications*, San Diego, CA, USA, 2003, pp. 115–121
2. M. Younis and K. Akkaya, 'Node positioning for increased dependability of wireless sensor networks', in *Algorithms and Protocols for Wireless Sensor Networks*, A. Boukerche (Ed.). Hoboken, NJ: John Wiley & Sons, 2009, pp. 225–266
3. B. Liu, O. Dousse, J. Wang and A. Saipulla, 'Strong barrier coverage of wireless sensor networks', in *The ACM International Symposium on Mobile Ad Hoc Networking and Computing*, Hong Kong, China, 2008, pp. 411–420
4. F. Pedraza, A.L. Medaglia and A. García, 'Efficient coverage algorithms for wireless sensor networks', in *IEEE Systems and Information Engineering Design Symposium*, Charlottesville, VA, USA, 2006, pp. 79–84
5. J. Lu, J. Wang and T. Suda, 'Scalable coverage maintenance for dense wireless sensor networks', *EURASIP Journal on Wireless Communications and Networking*, vol. 2007, p. 13, 2007
6. G. Xing, R. Tan, B. Liu, J. Wang, X. Jia and C. Yi, 'Data fusion improves the coverage of wireless sensor networks', in *International Conference on Mobile Computing and Networking*, Beijing, China, 2009, pp. 157–168
7. A. Fanimokun and J. Frolik, 'Effects of natural propagation environments on wireless sensor network coverage area', in *Proceedings of the 35th Southeastern Symposium on System Theory*, Morgantown, WV, USA, 2003, pp. 16–20
8. B. Liu, P. Brass, O. Dousse, P. Nain and D. Towsley, 'Mobility improves coverage of sensor networks', in *Proceedings of the 6th ACM International Symposium on Mobile Ad Hoc Networking and Computing*, Urbana-Champaign, IL, USA, 2005, pp. 300–308
9. S. Basagni, A. Carosi and C. Petrioli, 'Mobility in wireless sensor networks', in *Algorithms and Protocols for Wireless Sensor Networks*, A. Boukerche (Ed.). Hoboken, NJ: John Wiley & Sons, 2009, pp. 267–306
10. V.N. Christopoulos and S. Roumeliotis, 'Adaptive sensing for instantaneous gas release parameter estimation', in *IEEE International Conference on Robotics and Automation*, Barcelona, Spain, 2005, pp. 4450–4456
11. V. Hombal, A.C. Sanderson and R. Blidberg, 'A non-parametric iterative algorithm for adaptive sampling and robotic vehicle path planning', in *IEEE/*

*RSJ International Conference on Intelligent Robots and Systems*, Beijing, China, 2006, pp. 217–222
12. D.O. Popa, M.F. Mysorewala and F.L. Lewis, 'EKF-based adaptive sampling with mobile robotic sensor nodes', in *IEEE/RSJ International Conference on Intelligent Robots and Systems*, Beijing, China, 2006, pp. 2451–2456
13. D.O. Popa, A.C. Sanderson, R.J. Komerska, S.S. Mupparapu, D.R. Blidberg and S.G. Chappel, 'Adaptive sampling algorithms for multiple autonomous underwater vehicles', in *Autonomous Underwater Vehicles*, Sebasco Estates, ME, USA, 2004, pp. 108–118
14. A. Singh, D. Budzik, W. Chen, M.A. Batalin, M. Stealey, H. Borgstrom, *et al.*, 'Multiscale sensing: A new paradigm for actuated sensing of high frequency dynamic phenomena', in *IEEE/RSJ International Conference on Intelligent Robots and Systems*, Beijing, China, 2006, pp. 328–335
15. W. Jatmiko, K. Sekiyama and T. Fukuda, 'A mobile robots PSO-based for odor source localization in dynamic advection–diffusion environment', in *IEEE/RSJ International Conference on Intelligent Robots and Systems*, Daegu, South Korea, 2006, pp. 4527–4532
16. R.A. Russell, A. Bab-Hadiashar, R.L. Shepherd and G.G. Wallace, 'A comparison of reactive robot chemotaxis algorithms', *Robotics and Autonomous Systems*, vol. 45, pp. 83–97, November 2003
17. D.W. Scott, *Multivariate Density Estimation: Theory, Practice, and Visualization*. New York, NY, USA: Wiley-Interscience, 1992
18. B.W. Silverman, *Density Estimation for Statistics and Data Analysis*. London, UK: Chapman & Hall/CRC, 1986
19. A. Elfes, 'Using occupancy grids for mobile robot perception and navigation', *Science*, vol. 22, pp. 46–57, 1989
20. K.H. Low, G. Gordon, J. Dolan and P. Khosla, 'Adaptive sampling for multi-robot wide-area exploration', in *IEEE International Conference on Robotics and Automation*, Estoril, Portugal, 2007, pp. 755–760
21. D. Fox, W. Burgard, H. Kruppa and S. Thrun, 'A probabilistic approach to collaborative multi-robot localization', *Autonomous Robots*, vol. 8, pp. 325–344, June 2000
22. S.I. Roumeliotis and G.A. Bekey, 'Distributed multirobot localization', *IEEE Transactions on Robotics and Automation*, vol. 18, pp. 781–795, 2002
23. E.L. Creed, S.M. Glenn and R. Chant, 'Adaptive sampling experiment at LEO-15', in *OCC 1998 Proceedings, Marine Technology Section*, Baltimore, MD, USA, 1998, pp. 576–579
24. F.L. Lewis, L. Xie and D.O. Popa, *Optimal and Robust Estimation: With an Introduction to Stochastic Control Theory*. Boca Raton, FL, USA: CRC Press, 2007
25. N. Bulusu, J. Heidemann and D. Estrin, 'GPS-less low cost outdoor localization for very small devices', *IEEE Personal Communications Magazine*, vol. 7, pp. 28–34, 2005

26. J. Bachrach and C. Taylor, 'Localization in sensor networks', in *Handbook of Sensor Networks: Algorithms and Architectures*, 1st ed., vol. 1, I. Stojmenovi (Ed.). Hoboken, NJ, USA: Wiley, 2005
27. B. Hofmann-Wellenhof, H. Lichtenegger and J. Collins, *Global Positioning System: Theory and Practice*. New York: Springer, 2004
28. R.L. Moses, D. Krishnamurthy and R.M. Patterson, 'A self-localization method for wireless sensor networks', *EURASIP Journal on Applied Signal Processing*, vol. 4, pp. 348–358, 2002
29. N. Bulusu, J. Heidemann and D. Estrin, 'Adaptive beacon placement', in *ICDCS'01: Proceedings of the 21st International Conference on Distributed Computing Systems*, Phoenix, AZ, USA, 2001, p. 489
30. C. Savarese, J.M. Rabaey and J. Beutel, 'Locationing in distributed ad-hoc wireless sensor networks', in *IEEE International Conference on Acoustics, Speech, and Signal Processing*, Salt Lake City, UT, USA, 2001
31. N. Patwari, J.N. Ash, S. Kyperountas, A.O. Hero III, R.L. Moses and N.S. Correal, 'Locating the nodes: Cooperative localization in wireless sensor networks', *Signal Processing Magazine, IEEE*, vol. 22, pp. 54–69, July 2005
32. P. Bahl and V.N. Padmanabhan, 'RADAR: An in-building RF-based user location and tracking system', in *Proceedings of the IEEE Infocom 2000*, Tel Aviv, Israel, March 2000, pp. 775–784
33. N.B. Priyantha, A. Chakraborty and H. Balakrishnan, 'The cricket location-support system', in *Sixth Annual ACM International Conference on Mobile Computing and Networking (MOBICOM)*, Boston, MA, USA, August 2000
34. A. Savvides, C. Han and M.B. Strivastava, 'Dynamic fine-grained localization in ad-hoc networks of sensors', in *MobiCom'01: Proceedings of the 7th Annual International Conference on Mobile Computing and Networking*, Rome, Italy, July 2001, pp. 166–179
35. D. Niculescu and B. Nath, 'Ad hoc positioning system (APS) using AoA', in *Proceedings of INFOCOM*, San Francisco, CA, USA, March 2003
36. A. Boukerche, H.A.B.F. Oliveira, E.F. Nakamura and A.A.F. Loureiro, 'Localization systems for wireless sensor networks', in *Algorithms and Protocols for Wireless Sensor Networks*, A. Boukerche (Ed.). Hoboken, NJ: John Wiley & Sons, 2009, pp. 307–340
37. G. Mao and B. Fidan, *Localization Algorithms and Strategies for Wireless Sensor Networks*. Hershey, PA: Information Science Reference, 2009
38. L. Hu and D. Evans, 'Localization for mobile sensor networks', in *Proceedings of the 10th Annual International Conference on Mobile Computing and Networking*, Philadelphia, PA, USA, September 2004, pp. 45–57
39. C. Taylor, A. Rahimi, J. Bachrach and H. Shrobe, 'Simultaneous localization and tracking in an ad hoc sensor network', *Information Procession in Sensor Networks*, Nashville, TN, USA, 2005
40. A.M. Brooks, S. Williams and A. Makarenko, 'Automatic online localization of nodes in an active sensor network', *International Conference on Robotics and Automation*, vol. 5, pp. 4821–4826, 2004

41. S. Shenoy and J. Tan, 'Simultaneous localization and mobile robot navigation in a hybrid sensor network', in *IEEE/RSJ International Conference on Intelligent Robots and Systems*, Edmonton, Alberta, Canada, August 2005
42. A. Galstyan, B. Krishnamachari, K. Lerman and S. Pattem, 'Distributed online localization in sensor networks using a moving target', in *IPSN'04: Proceedings of the Third International Symposium on Information Processing in Sensor Networks*, Berkeley, CA, USA, 2004, pp. 61–70
43. V. Cevher and J.H. McClellan, 'Sensor array calibration via tracking with the extended Kalman filter', *IEEE International Conference on Acoustics, Speech, and Signal Processing*, vol. 5, pp. 2817–2820, 2001
44. P.N. Pathirana, N. Bulusu, A.V. Savkin and S. Jha, 'Node localization using mobile robots in delay-tolerant sensor networks', *IEEE Transactions on Mobile Computing*, vol. 4, pp. 285–296, May/June 2005
45. E. Olson, J. Leonard and S. Teller, 'Robust range-only beacon localization', in *IEEE/OES Autonomous Underwater Vehicles*, Sebasco Estates, ME, USA, 2004, pp. 66–75
46. W. Xiao, J.K. Wu and L. Xie, 'Sensor scheduling for target tracking in networks of active sensors', in *IEEE International Workshop on Sensor Networks and Applications*, Beijing, China, 2005
47. M.W.M.G. Dissanayake, P. Newman, S. Clark and H.F. Durrant-Whyte, 'A Solution to the Simultaneous Localization and Map Building (SLAM) Problem', *IEEE Transactions on Robotics and Automation*, vol. 17, June 2001, pp. 229–241
48. J.W. Fenwick, P.M. Newman and J.J. Leonard, 'Cooperative concurrent mapping and localization', in *Proceedings of the 2002 IEEE International Conference on Robotics and Automation*, Washington, DC, USA, May 2002, pp. 1810–1817
49. S. Thrun, W. Burgard and D. Fox, *Probabilistic Robotics*. Cambridge, MA: MIT Press, 2005
50. D.O. Popa, K. Sreenath and F.L. Lewis, 'Robotic deployment for environmental sampling applications', in *International Conference on Control and Automation*, Budapest, Hungary, 2005, pp. 197–202
51. F.L. Lewis, D. Tacconi, A. Gurel, H.H. Huang and O.C. Pastravanu, 'Manufacturing controller design and deadlock avoidance using a matrix model for discrete event systems', in *Methods and Applications of Intelligent Control*, 1st ed., S.G. Tzafestas (Ed.). Netherlands: Kluwer Academic Publishers, 1997, pp. 485–508
52. J.G. Mireles Jr., 'Matrix-Based Intelligent Discrete Event Control for Flexible Manufacturing Systems', Ph.D. Dissertation/Thesis, The University of Texas at Arlington, Arlington, TX, May 2002
53. J. Mireles Jr., F.L. Lewis, A. Gurel and S. Bogdan, 'Deadlock avoidance algorithms and implementation, a matrix based approach', in *Deadlock Resolution in Computer-Integrated Systems*, 1st ed., vol. 1, M. Zhou and M.P. Fanti (Eds.). New York, NY: Marcel Dekker, 2004, pp. 183–232

54. T. Cheng, 'On-line deadlock avoidance for complex routing flexible manufacturing cells', *International Journal of Applied Science and Engineering*, vol. 2, pp. 163–176, 2004
55. H.J. Yoon and D.Y. Lee, 'Deadlock-free scheduling of photolithography equipment in semiconductor fabrication', *IEEE Transactions on Semiconductor Manufacturing*, vol. 17, pp. 42–54, February 2004
56. M. Koibuchi, A. Jouraku, K. Watanabe and H. Amano, 'Descending layers routing: A deadlock-free deterministic routing using virtual channels in system area networks with irregular topologies', in *Proceedings of 2003 International Conference on Parallel Processing*, Kaohsiung, Taiwan, 2003, pp. 527–536
57. V. Giordano, F.L. Lewis, J. Mireles Jr. and B. Turchiano, 'Coordination control policy for mobile sensor networks with shared heterogeneous resources', in *2005 International Conference on Control and Automation*, Budapest, Hungary, 2005, pp. 191–196
58. J. Ezpeleta, J.M. Colom and J. Martinez, 'A Petri net based deadlock prevention policy for flexible manufacturing systems', *IEEE Transactions on Robotics and Automation*, vol. 11, pp. 173–184, April 1995
59. Y. Huang, M. Jeng, X. Xie and S. Chung, 'Deadlock prevention policy based on Petri nets and siphons', *International Journal of Production Research*, vol. 39, pp. 283–305, 2001
60. K. Lautenbach, 'Linear algebraic calculation of deadlocks and traps', in *Concurrency and Nets*, 1st ed., vol. 1, K. Voss, H.J. Genrich and G. Rozenberg (Eds.). Berlin, Germany: Springer-Verlag, 1987, pp. 315–336
61. G. Alpan and B. Gaujal, 'Supervisory control of Petri nets using routing functions: Starvation avoidance issues', *IEEE Transactions on Systems, Man and Cybernetics, Part B*, vol. 30, pp. 684–694, October 2000
62. F.L. Lewis, S. Bogdan, A. Gurel and O. Pastravanu, 'Analysis of deadlocks and circular waits using a matrix model for discrete event systems', in *Proceedings of the 36th IEEE Conference on Decision and Control*, San Diego, CA, USA, 1997, pp. 4080–4085
63. F.L. Lewis, A. Gurel, S. Bogdan, A. Doganalp and O.C. Pastravanu, 'Analysis of deadlock and circular waits using a matrix model for flexible manufacturing systems', *Automatica*, vol. 34, pp. 1083–1100, 1998
64. J. Mireles Jr., F.L. Lewis and A. Gurel, 'Deadlock avoidance for manufacturing multipart re-entrant flow lines using a matrix-based discrete event controller', *International Journal of Production Research*, vol. 40, pp. 3139–3166, 2002
65. M. Lawley, 'Integrating flexible routing and algebraic deadlock avoidance policies in automated manufacturing systems', *International Journal of Production Research*, vol. 38, pp. 2931–2950, 2000
66. M. Lawley, 'Flexible routing and deadlock avoidance in automated manufacturing systems', in *Proceedings of the 1998 IEEE International Conference on Robotics and Automation*, Leuven, Belgium, 1998, pp. 591–596

67. J. Mireles Jr. and F.L. Lewis, 'Deadlock analysis and routing on free-choice multipart reentrant flow lines using a matrix-based discrete event controller', in *Proceedings of the 41st IEEE Conference on Decision and Control*, Las Vegas, NV, USA, 2002, pp. 793–798
68. J. Borenstein and L. Feng, 'Measurement and correction of systematic odometry errors in mobile robots', *IEEE Transactions on Robotics and Automation*, vol. 12, pp. 869–880, 1996
69. J. Borenstein and L. Feng, 'UMBmark – A Method for Measuring, Comparing, and Correcting Odometry Errors in Mobile Robots', Tech. Rep. UM-MEAM-94-22, The University of Michigan, Ann Arbor, MI, December 1994
70. J. Borenstein and L. Feng, 'UMBmark: A benchmark test for measuring odometry errors in mobile robots', in *Proceedings of the SPIE Conference on Mobile Robots*, Philadephia, PA, USA, 1995
71. S.K. Thompson, *Sampling*. New York, NY: Wiley-Interscience, 2002
72. F. van der Heijden, R. Duin, D. de Ridder and D.M.J. Tax, *Classification, Parameter Estimation and State Estimation: An Engineering Approach Using MATLAB*. London, UK: John Wiley & Sons, 2004
73. O. Nelles, *Nonlinear System Identification: From Classical Approaches to Neural Networks and Fuzzy Models*. Germany: Springer, 2000
74. J.C. Bezdek, *Pattern Recognition with Fuzzy Objective Function Algorithms*. Norwell, MA: Kluwer Academic Publishers, 1981
75. J. Park and I.W. Sandberg, 'Universal approximation using radial-basis-function networks', *Neural Computation*, vol. 3, pp. 246–257, 1991
76. R. Willett, A. Martin and R. Nowak, 'Backcasting: Adaptive sampling for sensor networks', in *IPSN'04: Proceedings of the Third International Symposium on Information Processing in Sensor Networks*, Berkeley, CA, USA, 2004, pp. 124–133
77. A. Howard, M.J. Mataric and G.S. Sukhatme, 'Mobile sensor network deployment using potential fields: A distributed, scalable solution to the area coverage problem', in *Proceedings of the 6th International Symposium on Distributed Autonomous Robotics Systems*, Fukuoka, Japan, 2002, pp. 299–308
78. J.A. Farrell, S. Pang and W. Li, 'Chemical plume tracing via an autonomous underwater vehicle', *IEEE Journal of Oceanic Engineering*, vol. 30, pp. 428–442, April 2005
79. R.A. Russell, *Odour Sensing for Mobile Robots*. Singapore/New Jersey/London/Hong Kong: World Scientific, 1999
80. M.Y. Rachkov, L. Marques and A.T. de Almeida, 'Multisensor demining robot', *Autonomous Robots*, vol. 18, pp. 275–291, May 2005
81. H. Ishida, T. Ushiku, S. Toyama, H. Taniguchi and T. Moriizumi, 'Mobile robot path planning using vision and olfaction to search for a gas source', in *IEEE Sensors*, Irvine, CA, USA, 2005, pp. 1112–1115
82. W. Jatmiko, Y. Ikemoto, T. Matsuno, T. Fukuda and K. Sekiyama, 'Distributed odor source localization in dynamic environment', in *IEEE Sensors*, Irvine, CA, USA, 2005, pp. 254–257

83. V.N. Christopoulos and S. Roumeliotis, 'Multi robot trajectory generation for single source explosion parameter estimation', in *IEEE International Conference on Robotics and Automation*, Barcelona, Spain, 2005, pp. 2803–2809
84. A.F. Bennett, *Inverse Modeling of the Ocean and Atmosphere*. Cambridge, UK: Cambridge University Press, 2002
85. D.O. Popa, H.E. Stephanou, C. Helm and A.C. Sanderson, 'Robotic deployment of sensor networks using potential fields', in *IEEE International Conference on Robotics and Automation*, New Orleans, LA, USA, 2004, pp. 642–647
86. C.J. Cannell and D.J. Stilwell, 'A comparison of two approaches for adaptive sampling of environmental processes using autonomous underwater vehicles', in *Proceedings of MTS/IEEE OCEANS*, Washington, DC, USA, 2005, pp. 1514–1521
87. K. Sreenath, 'Adaptive Sampling with Mobile WSN', M.S. Dissertation/Thesis, The University of Texas at Arlington, Arlington, TX, December 2005
88. V.V. Fedorov, W.J. Studden and E.M. Klimko, *Theory of Optimal Experiments*. USA: Academic Press Inc., New York, NY, USA, 1972
89. E. Chuvieco (Ed.), *Wildland Fire Danger Estimation and Mapping; The Role of Remote Sensing Data*. Singapore: World Scientific Publishing Company, 2004
90. M.A. Finney, 'FARSITE: Fire Area Simulator – Model Development and Evaluation', Tech. Rep. RMRS-RP-4, U.S. Department of Agriculture, Forest Service, Rocky Mountain Research Station, Ogden, UT, 1998
91. E.A. Johnson and K. Miyanishi (Eds.), *Forest Fires: Behavior and Ecological Effects*. San Diego, CA: Academic Press, 2001
92. R.C. Rothermel, 'How to Predict the Spread and Intensity of Forest and Range Fires', Tech. Rep. INT-143, U.S. Department of Agriculture, Forest Service, Intermountain Forest and Range Experiment Station, Ogden, UT, 1983
93. I. Karafyllidis and A. Thanailakis, 'A model for predicting forest fire spreading using cellular automata', *Ecological Modelling*, vol. 99, pp. 87–97, 1997
94. X. Li and A.G. Yeh, 'Neural-network-based cellular automata for simulating multiple land use changes using GIS', *International Journal of Geographical Information Science*, vol. 16, pp. 323–343, 2002
95. A. Muzy, E. Innocenti, A. Aiello, J. Santucci and G. Wainer, 'Specification of discrete event models for fire spreading', *Simulation*, vol. 81, pp. 103–117, 2005
96. G.A. Trunfio, 'Predicting wildfire spreading through a hexagonal cellular automata model', in *ACRI 2004: International Conference on Cellular Automata for Research and Industry*, Amsterdam, The Netherlands, 2004, pp. 385–394
97. S.S. Ge and Y.J. Cui, 'Dynamic motion planning for mobile robots using potential field method', *International Journal of Autonomous Robots*, vol. 13, pp. 207–222, 2002
98. J. Latombe, *Robot Motion Planning*. Boston, MA: Kluwer Academic, 1991

99. B.H. Krogh, 'A generalized potential field approach to obstacle avoidance control', in *International Robotics Research Conference*, Bethlehem, PA, USA, 1984, pp. 1150–1156
100. O. Kathib, 'Real-time obstacle avoidance for manipulators and mobile robots', in *IEEE International Conference on Robotics and Automation*, St. Louis, MO, USA, 1985, pp. 500–505
101. S. Huang and G. Dissanayake, 'Convergence and consistency analysis for extended Kalman filter based SLAM', *IEEE Transactions on Robotics*, vol. 23, pp. 1036–1049, 2007
102. B.S. Rao and H.F. Durrant-Whyte, 'Fully decentralized algorithm for multisensor Kalman filtering', *IEE Proceedings on Control Theory and Applications*, vol. 138, pp. 413–420, 1991
103. J.P. Desai, J.P. Ostrowski and V. Kumar, 'Modeling and control of formations of non-holonomic mobile robots', *IEEE Transactions on Robotics and Automation*, vol. 17, pp. 905–908, 2001
104. Y. Gao, E.Y. Krakiwsky, M.A. Abousalem and J.F. Mclellan, 'Comparison and analysis of centralized, decentralized, and federated filters', *Journal of the Institute of Navigation*, vol. 40, pp. 69–86, 1993
105. Q. Du, V. Faber and M. Gunzburger, 'Centroidal Voronoi tessellations: Applications and algorithms', *SIAM Review*, vol. 41, pp. 637–676, 1999
106. R. Olfati-Saber, 'Distributed Kalman filter with embedded consensus filters', in *44th IEEE Conference on Decision and Control, 2005 and 2005 European Control Conference*, Seville, Spain, 2005, pp. 8179–8184
107. F.L. Lewis, D.A. Tacconi, O.C. Pastravanu and A. Gurel, 'Method and Apparatus for Testing and Controlling a Flexible Manufacturing System', U.S. Patent 6,185,469, 6 February 2001
108. D.A. Tacconi and F.L. Lewis, 'A new matrix model for discrete event systems: Application to simulation', *IEEE Control Systems Magazine*, vol. 17, pp. 62–71, October 1997
109. V. Giordano, F.L. Lewis, B. Turchiano, P. Ballal and V. Yeshala, 'Matrix computational framework for discrete event control of wireless sensor networks with some mobile agents', in *Proceedings of Mediterranean Conference on Control and Automation*, Limassol, Cyprus, 2005
110. P. Ballal, F.L. Lewis, J. Mireles Jr. and K. Sreenath, 'Deadlock avoidance for free choice multi-reentrant flow lines: Critical siphons & critical subsystems', in *IEEE Mediterranean Conference on Control and Automation*, Athens, Greece, 2007, pp. 1–8
111. P.S. Maybeck, *Stochastic Models, Estimation, and Control*. USA: Academic Press, New York, NY, USA, 1979
112. W.S. Levine (Ed.), *The Control Handbook*. New York, NY: CRC Press, 1996
113. F.L. Lewis, *Optimal Estimation*. New York, NY: John Wiley & Sons, 1986
114. S.I. Roumeliotis and G.A. Bekey, 'An extended Kalman filter for frequent local and infrequent global sensor data fusion', in *SPIE International*

*Symposium on Intelligent Systems and Advanced Manufacturing*, Pittsburgh, PA, USA, October 1997, pp. 11–22

115. E. Kiriy and M. Buehler, 'Three-state Extended Kalman Filter for Mobile Robot Localization', Tech. Rep. TR-CIM 05.06, McGill University, Montreal, Canada, April 2002
116. A. Kelly, 'A 3D State Space Formulation of a Navigation Kalman Filter for Autonomous Vehicles', Tech. Rep. CMU-RI-TR-94-19, Carnegie Mellon University, Pittsburgh, PA, May 1994
117. D. Fox, W. Burgard, F. Dellaert and S. Thrun, 'Monte Carlo localization: Efficient position estimation for mobile robots', in *Proceedings of the Sixteenth National Conference on Artificial Intelligence (AAAI'99)*, Orlando, FL, USA, July 1999
118. E.W. Weisstein, *Frobenius Norm. From MathWorld – A Wolfram Web Resource*. Available at http://mathworld.wolfram.com/FrobeniusNorm.html
119. G. Zhou, T. He, S. Krishnamurthy and J.A. Stankovic, 'Impact of radio irregularity on wireless sensor networks', in *International Conference on Mobile Systems, Applications and Services*, Boston, MA, USA, 2004, pp. 125–138
120. G. Zhou, T. He, S. Krishnamurthy and J.A. Stankovic, 'Models and solutions for radio irregularity in wireless sensor networks', *ACM Transactions on Sensor Networks*, vol. 2, pp. 221–262, May 2006
121. A. Woo, T. Tong and D. Culler, 'Taming the underlying challenges of reliable multihop routing in sensor networks', in *Proceedings of the 1st International Conference on Embedded Networked Sensor Systems*, Los Angeles, CA, USA, 2003, pp. 14–27
122. G. Lu, N. Sadagopan, B. Krishnamachari and A. Goel, 'Delay efficient sleep scheduling in wireless sensor networks', in *Proceedings of the 24th Annual Joint Conference of the IEEE Computer and Communication Societies*, Miami, FL, USA, March 2005, pp. 2470–2481
123. L. Wang and Y. Xiao, 'A survey of energy-efficient scheduling mechanisms in sensor networks', *Mobile Networks and Applications*, vol. 11, pp. 723–740, October 2006
124. W. Ye, J. Heidemann and E. Deborah, 'An energy-efficient MAC protocol for wireless sensor networks', in *Proceedings of the IEEE Infocom*, New York, NY, USA, 2002, pp. 1567–1576
125. T.V. Dam and K. Langendoen, 'An adaptive energy-efficient MAC protocol for wireless sensor networks', in *Proceedings of the 1st International Conference on Embedded Networked Sensor Systems*, Los Angeles, CA, USA, 2003, pp. 171–180

# Index